SpringerBriefs in Applied Sciences and Technology

Computational Intelligence

Series Editor

Janusz Kacprzyk, Systems Research Institute, Polish Academy of Sciences, Warsaw, Poland

SpringerBriefs in Computational Intelligence are a series of slim high-quality publications encompassing the entire spectrum of Computational Intelligence. Featuring compact volumes of 50 to 125 pages (approximately 20,000-45,000 words), Briefs are shorter than a conventional book but longer than a journal article. Thus Briefs serve as timely, concise tools for students, researchers, and professionals.

Soni Sweta

Sentiment Analysis and its Application in Educational Data Mining

 Springer

Soni Sweta
Department of Computer Engineering,
Mukesh Patel School of Technology
Management & Engineering
SVKM's Narsee Monjee Institute
of Management Studies (NMIMS)
Deemed-to-University
Mumbai, Maharashtra, India

ISSN 2191-530X   ISSN 2191-5318 (electronic)
SpringerBriefs in Applied Sciences and Technology
ISSN 2625-3704   ISSN 2625-3712 (electronic)
SpringerBriefs in Computational Intelligence
ISBN 978-981-97-2473-4   ISBN 978-981-97-2474-1 (eBook)
https://doi.org/10.1007/978-981-97-2474-1

This Springer imprint is published by the registered company Springer Nature Singapore Pte Ltd.
The registered company address is: 152 Beach Road, #21-01/04 Gateway East, Singapore 189721,
Singapore

Paper in this product is recyclable.

Late Shree Dhirendra Pratap Singh

*Your Inspiring Words to Work Hard and
Divine Blessings "Mahamangal Narayan"...
This Book is dedicated to my Father-in-law
Late Shree Dhirendra Pratap Singh*

Dr. Soni Sweta

Foreword

Education is essential for all human beings which is a life-long journey. It is desired by the governments and academicians to impart the best education in the most effective and adaptable way to upgrade learning skills. Sentiment analysis is the new emerging dimension which is well stressed upon by the author, i.e. the exploration of sentiments hidden in the interactive educational processes across the various educational systems from the traditional one to artificial intelligence driven system.

Understanding the integration of sentiments incorporated into educational processes reveals multiple and valuable insights useful for all the stakeholders participating in the entire learning process. Showcasing the significance of sentiment analysis over Educational Data Mining is the theme of this book. This book *Sentiment Analysis and its Application in Educational Data Mining* is guiding compass to educators, researchers, and technocrats. It explores comprehensively about the sentiment analysis and its application in diverse areas. It further describes tools and techniques used, integration of sentiment analysis with Educational Data Mining, and in last the emerging trends and challenges.

Apart from study of enormous data and their patterns, the emotional factors took centre stage of educational environment in view of new emerging technologies and learner-centric empathetic analysis. This book opens a beacon of hope to decipher emotional aspects of the educational interactions. It helped to start a new wave of discussion in line of building very strong educational ecosystem.

This book embarks a new journey of rejuvenating the sentiments and consolidating the aspirations of millions of stakeholders to factor emotional dimensions before arriving at any final decisions. The enlightened benefits of the sentiments may be augmented in new age educational pursuits.

The author of this book Dr. Soni Sweta is well known to me since long. I always encouraged and guided her for achieving many laurels in her career and witnessed her successes through her transformational journey. I appreciate her for exploring such a challenging dimension of the educational aspects which have created a new thought process to understand and implement them effectively in the education system.

I am very much glad to share some important things about the book chapters; first two chapters Chaps. 1 and 2 discuss the overview of Sentiment Analysis and its Application in the diverse domains. Chapters 3 and 4 have described the impact of sentiment analysis over the education sector and technology used in sentiment analysis with integration of Educational Data Mining which strengthens the core idea of this book and makes this book as incredible in terms of significance and utility. The remaining chapters give a wide view of the tool, methods, algorithm, and implementation, emerging trends and challenges of sentiment analysis in the domain of the education systems.

She has done wonderful job to sum up the sentiments analysis and discuss the future scope of research and studies in this area of interest.

Dr. Soni Sweta has more than 20 years of teaching experiences. So, I think she has enough knowledge in field of education system and the core subject areas discussed in the book. I appreciate her to put her experience and knowledge in form of book to give a huge benefit to the society.

I think this is a great contribution by author in her teaching career. I wish her all the best for future, new editions, and many more such interesting books awaiting to unfold from her side.

Prof. (Dr.) D. K. Singh
Vice Chancellor
Jharkhand University of Technology
Ranchi, India

Preface

Dear Readers,

Let me welcome and introduce to the exciting journey of my book *Sentiment Analysis and its Applications in Educational Data Mining*. I am delighted to navigate you through this amazing world that may satisfy your inquisitive mind to understand the link between Sentiment Analysis and its wide applications in the landscape of Educational Data Mining.

In today's dynamic world, understanding the hidden emotional patterns is key behind the learner's success. The blending and fusion of sentiment analysis augmented a powerful tool to decode sentiments and enhance learning outcomes. It is an innovative approach to extract such knowledge from EDM to uphold the future promises. It will transform the entire landscape of educational system. Engaging the influential factors of sentiments, we can optimize learning experience.

This book, *Sentiment Analysis and its Applications in Educational Data Mining*, aims to resolve the core interconnecting factors of the sentiment analysis and educational data mining and how the learnings can be applied in real-life situations. As an author, I shared my research experiences, codified important witnesses and depicted profound learnings with insightful analysis. My motivation stems from witnessing the huge impact that insightful analysis is not only to help academicians to make educational strategies, but also to foster and inculcate the habits of using sentimental prowess for better learning outcomes.

An outline for Exploration and Navigation- Embark your journey:

Navigating the terrain starts from Chap. 1 "An Overview of Sentiment Analysis and Educational Data Mining."

The journey begins with understanding the fundamental of sentimental analysis, its types, role, and factors to analysis of sentiments along with Educational Data Mining and Its concepts.

Chapter 2 "Application of Sentiment Analysis in Diverse Domains" explore the advantages of sentiment analysis task in different domain. This chapter illustrates a study on various applications of sentiment analysis that is from marketing and business to finance, commerce and trade, health care to pharma, aviation, and beyond. It showcased how the sentiment analysis adaptively predicting the needs and recommending best solutions for different industries.

It covers the applications and uses in all important areas including educational domain and its benefits of applying it across the various domains.

Sentiment Analysis revolutionizing and speeding up entire landscape as a catalyst described in Chap. 3 "The Transformative Role of Sentiment Analysis in Education."

This book chapter analyses multifaceted applications of sentiment analysis in education, explores its significant impact on educational approaches, discusses learner's engagement, and helps in formulating institutional strategies. The chapter describes all about the through key themes, including the importance of the sentiment analysis in assessing learner's feedback, the improvement of their personalized learning experiences and the overall development of a positive learning spaces. This chapter describes its relevance and effectiveness in education system. It helps to improve the overall effectiveness of the education system. It explains insights into the transformative potential of integration of sentiment analysis tools in unhiding the complex tapestry of human emotions within the digital world.

About tools and techniques for sentiment analysis discussed in Chap. 4 "Sentiment Tech: Exploring the Tools Shaping Emotional Analysis."

This chapter illustrates an in-depth study about the contemporary sentiment analysis tools and techniques and algorithm across different approaches and elaborates how researchers, administrator, educators, and other stakeholders using this capability to extract significant hidden insights from large datasets for comprehensive understanding. From natural language processing algorithms to machine learning models, the chapter depicts the diverse tools available for capturing and studying sentiments, whether positive, negative, or neutral. With a focus on real-world applications, the chapter discusses the functionalities of the tools used for social media monitoring, customer feedback analysis, and beyond. This chapter unfolds the threads of lexicon-based approaches, machine learning techniques, and the burgeoning realm of deep learning and natural language processing. It identifies the key methodologies, models, and algorithms applied in the analysis of sentiments expressed during the learning processed and captured in educational data at the time of occurrence.

Comprehensive exploration on the evolving landscape posed some challenges discussed in Chap. 5 "Emerging Trends and Challenges in Educational Sentiment Analysis."

This book chapter serves as a valuable resource for educator, researcher, and scholars seeking to grasp the current trends of educational sentiment analysis. Multimodal sentiment analysis, context-aware methodologies, and the integration of explainable AI are among the most important trends explained in this chapter. The chapter examines the current challenges that practitioners and researchers encounter in educational sentiment analysis. Interpretability, ethical considerations, and the demand for context-aware analyses discussed as critical challenges and how to address them are also described.

Unveiling the Impact: One of the book's defining features is to showcase the critical impact of sentiment analysis on educational outcomes. From adaptive learning systems to personalized interventions, we uncover valuable insights derived from sentiments helping decision-making processes, enriching the educational journey for both educators and learners.

Pioneering the Future, Beyond the Horizon: This book doesn't merely sojourn on the present but also dwells about the future research related to Sentiment Analysis in Educational Data Mining. We contemplate the uncharted territories of cross-domain sentiment analysis, multilingual sentiment understanding, and the integration of sentiment analysis with other educational analytics techniques. In this chapter, we aim to provide a compass for educators, researchers, and technologists navigating the evolving ideas of educational sentiment analysis.

A Call to Action: Educational sentiment analysis isn't just a technical and digital pursuit; it's platform to create learning environments that resonate with the emotions of every learner. As we unveil this book, I invite you to embark on this intellectual journey, armed with curiosity and a commitment to harnessing the potential of sentiment analysis with future trends.

Dear readers, as you immerse yourselves in the pages of *Sentiment Analysis and its Applications in Educational Data Mining*, I encourage you not just to passively read but to envision your role in this narrative. Whether you are an educator, a student, a researcher, academician, or a technology enthusiast, your perspectives and insights contribute to the evolving story of educational sentiment analysis.

The book concludes by summarizing the key findings and insights presented throughout the text. It also offers the future of sentiment analysis in Educational Data Mining (EDM), highlighting the potential of these techniques to revolutionize educational research and practice.

This book is intended for researchers, educators, and practitioners who are interested in applying sentiment analysis techniques to improve educational outcomes. It provides a comprehensive introduction to the field and offers valuable insights into the practical applications of these techniques in EDM.

May this book inspire you, spark new ideas, and empower you to take the benefits of the transformative potential of sentiment analysis in the realm of education.

Happy reading!

Mumbai, India   Dr. Soni Sweta

Acknowledgements

First and foremost, I would like to put my sincere gratitude to Chancellor **Shri. Amrish Patel**, Vice-Chancellor **Dr. Ramesh Bhat**, Pro Vice-Chancellor **Dr. Sharad Mhaiskar**, whose guiding visions and leading academic pursuits are behind the synthesis of this book. Your overall support for providing available resources and believing in my abilities to author the book.

I extent my sincere thanks to Dean **Dr. Alka Mahajan**, Dean Research **Prof. Archana Bhise**, HoD **Prof. Dhirendra Misra** for your inspiring guidance, true orientation, and continual encouragement during writing of the book. Your valuable inputs and effective mentoring have paved the way for this work of creativity.

I express my special thanks to my **all-fellow academicians**, working in the CE, MPSTME, NMIMS, for your numerous contributions like meaningful insights and constructive feedback, which have remarkably enriched this work. Your support has been a source of motivation.

I would also like to add sincere thanks to my **colleagues and friends** for their exchanging positive attributes, providing conducive environment and engaging in intellectual discussions which have helped in composition of this work. Their collaborative skills, genuine feedback, and practical suggestions had worked as lighthouse in view of improvement of this work.

I am truly indebted to my academician parents, **Prof. (Dr.) Shiv Kumari Singh** and **Late Prof. (Dr.) C. D. Roy**, for their unconditional love, encouragement and support. They have always promoted me and stand by my side. I will remain grateful for their sacrifices for my better upbringings. They are my pillars of strength.

I warmly appreciate and put my deepest gratitude to my husband **Mr. Daya Shankar** for his unswerving support, my in-laws, and all my family members for their mutual understanding and generous succour during the topsy-turvy paths of concoction of this book. Your love and affection have empowered me to triumph this journey.

I am also very much grateful to my son, **Mr. Arunabh Kshitij**, and my pet **Maggie** for their love and fun during this creative time. Their patience and making peaceful environment during peak of this focused work are highly commendable. Your cheerfulness and jovial delights brought meaning to the life and work. I am grateful to them from bottom of my heart for their radiant buoyancy. **Their presence in my life has always been a constant source of energy, strength, and inspiration, and their sacrifices have allowed me to dedicate the time and space obligatory to complete this synthesis**.

In last but not least, my sincere thanks to every individual who contributed in various capacities during this expedition. Your kind and sincere support have been the motivating force behind the success of this work.

Thanks to **almighty God** giving me strength and confidence to complete this book on time.

Dr. Soni Sweta

Contents

About the Author

Dr. Soni Sweta, Ph.D. in Computer Science and Engineering from BIT, Mesra, Ranchi, is presently working as an assistant professor in the Department of Computer Engineering at Mukesh Patel School of Technology Management & Engineering, NMIMS, Mumbai Campus, Mumbai, Maharashtra. She received her Master of Technology degree from Rajiv Gandhi Proudyogiki Vishwavidyalaya, Bhopal. She is presently guiding many Ph.D., M.Tech. M.C.A, and B.Tech. Scholars. Previously, she had guided many M.Tech., M.C.A, B.C.A, and B.Tech. students in their dissertation and final project work. Her present areas of research are artificial intelligence, natural language processing, soft computing, data mining, machine learning, data science, etc. Being a member of IEEE and life member of CSI (India), she is associated with few reputed journals as a reviewer and member of editorial board. She has acted as a technical committee member in many reputed conferences so far.

Abbreviations

ALS	Adaptive Learning Systems
ANN	Artificial Neural Network
BERT	Bidirectional Encoder Representations from Transformers
CI	Computational Intelligence
DL	Deep Learning
DM	Data Mining
EDM	Educational Data Mining
HCI	Human–Computer Interface
HR	Human Resources
IT	Information Technology
LS	Learning Style
LSTM	Long Short-Term Memory
ML	Machine Learning
RNNs	Recurrent Neural Networks
SA	Sentiment Analysis

Chapter 1
An Overview of Sentiment Analysis and Educational Data Mining

1.1 Introduction to Sentiment Analysis

In Sentiment Analysis, which is also known as opinion mining, the focus is on understanding how people feel about certain things including goods, services, organizations, people, issues, events, and themes. Subjectivity analysis, analysis of an effect, emotion, review mining, sentiment mining, opinion mining, opinion extraction, etc. all belong to the same Sentiment Analysis. Some of the terminology used to represent it as appraisal extraction and subjectivity analysis [1, 2]. Moreover, it has certain similarities to affective computing, which involves computer-based emotion identification and representation [3–5]. Subjective features are commonly examined in this topic, as defined by [6] "linguistic expressions of private states in context." These consist of singular words, phrases, or sentences at most. While there have been instances where whole documents have been analysed as sentiment units [7], the conclusion is more commonly present in smaller linguistic units. Because feeling and opinion often pertain to the same notion, throughout this study they will be utilized interchangeably. Sentiment in the text is classified into two types: explicit, where the subjective sentence communicates an opinion directly ("It's a beautiful day"), and implicit, where the language implies an opinion ("The earphone broke in two days"). Most of the study done thus far has been on the first type of sentiment since it is the easiest to analyse.

The text's polarity of emotions is a distinguishing quality. Polarity is normally divided into two categories: positive and negative, although it can also be conceived of as a range. A document with multiple subjective claims would have a mixed polarity overall, as opposed to having no polarity at all (being objective). Furthermore, a distinction must be established between sentiment's polarity and its strength. One may have strong feelings about a product being okay, not exceptionally good or bad, or weak feelings about a product being good (possibly because one owned it for too short a period to establish a strong opinion).

S. Sweta, *Sentiment Analysis and its Application in Educational Data Mining*,
SpringerBriefs in Computational Intelligence,
https://doi.org/10.1007/978-981-97-2474-1_1

Another crucial aspect of the sentiment is the target audience—an object, a concept, a person, or anything. A significant portion of the work has been dedicated to reviewing products and films, in which the underlying theme of the content is clear. However, it is frequently beneficial to focus on the aspect to which the author is referring: are purchasers more concerned with the battery life or the camera display? In the past decade, feature extraction has garnered considerable interest due to the accessibility of product review datasets. The mentioned attributes may be explicitly referenced ("Battery life is too brief") or subtly implied ("Camera is too big") within the text. Considerable data is derived from the sources.

Emotional Tone: A key challenge in Sentiment Analysis involves properly categorizing the personal opinions of the individuals that are being investigated. The concept of subjectivity was originally defined by linguists, most notably Randolph Quirk [3, 8]. According to Quirk, a private state cannot be objectively observed or verified. These private states comprise speculation sentiments, and opinions, among other elements. The notion of a private state essentially suggests that doing Sentiment Analysis will be challenging. Subjectivity is frequently referenced in conversation; it is highly dependent on the surrounding conditions, and its manifestation is typically unique to everyone. On the other hand, subjectivity does not always mean that anything is prone to errors or mistakes [6]. While the statement "Mary loves chocolate" perfectly conveys Mary's affection for chocolate, its truthfulness remains unchallenged. Similarly, not every declarative phrase is true.

To underscore the lack of clarity in the concept, Pang and Lee [9] provide the following definitions of concepts that are closely linked to sentiment:

- Opinion denotes a well-thought-out yet debatable conclusion that have the meanings of terms closely related to the concept of sentiment ("each expert seemed to have a different opinion").
- View suggests a subjective opinion ("very assertive in stating his views").
- Conviction applies to a party's firmly and seriously held belief ("the conviction that animal life is as sacred as human").
- Belief implies often deliberate acceptance and intellectual assent ("a firm belief in her party's platform").
- Persuasion suggests a belief grounded on assurance (as by evidence) of its truth ("was of the persuasion that everything changes").
- Sentiment suggests a settled opinion reflective of one's feelings ("her feminist sentiments are well-known").

Prominent natural language processing (NLP) researcher Wiebe [6] analysed point of view in narratives using Quirk's idea of the private state. The author provides a definition of private state as a tuple consisting of an experiencer's state p and an object's attitude (p, experiencer, attitude, object). Practically, a simplified version of this model is often used, wherein the polarity and the object of the emotion are the only factors considered. Several researchers provide a wide definition of sentiment as a positive or negative sentiment.

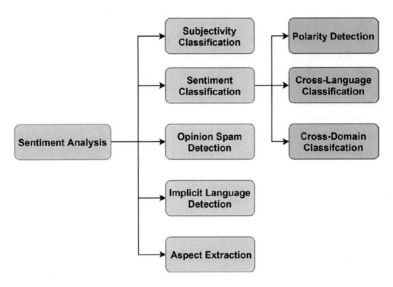

Fig. 1.1 Task of Sentiment Analysis [7]

Additionally, sentiment possesses a few qualities that set it apart from other textual elements that we would like to monitor. Frequently, we want content to be classified by subject, which may necessitate managing entire taxonomies of subjects. In contrast, sentiment classification often operates with a binary value (positive or negative), a range of polarity (e.g. ratings of movie stars), or even a range of opinion strength [5]. These classes cover a wide range of topics, individuals, and file formats. While it may seem like handling a few classes is a less complex undertaking than typical text analysis, nothing could be further from the truth.

Figure 1.1 shows the some of the tasks performed by Sentiment Analysis.

1.2 Different Types of Sentiment Analysis

- Binary Sentiment Analysis:
  Description: There are two types of text classification: positive and negative.
  Example: Identifying the sentiment of product review whether product review expresses positive or negative sentiment.
- Multiclass Sentiment Analysis:
  Description: Classifying text into multiple sentiment categories, such as various degrees of positivity/negativity or positive, negative, or neutral.
  Example: Categorizing customer reviews into positive, negative, or neutral sentiments.
- Aspect-Based Sentiment Analysis [10, 11]:

Description: In contrast to just detecting the overall sentiment of the content, aspect-based Sentiment Analysis seeks to uncover sentiments associated with specific elements or aspects stated in the text. Illustratively, concerns regarding "ambiance," "food," and "service" are identified and mitigated independently in a restaurant review. Analysing sentiment at a more granular level by evaluating certain characteristics or aspects inside the text.

Example: Assessing product reviews to identify sentiments related to various features such as cost, usability, and customer support.

- Emotion-Based Sentiment Analysis [3, 12]:
  Description: Emotion-based Sentiment Analysis is the determination and analysis of certain emotions expressed in a provided text. This approach outperforms traditional Sentiment Analysis methods by accurately capturing intricate emotional states. "Identify emotions such as happiness, sorrow, rage, or amazement in social media posts." Recognizing emotions that are expressed in the text, such as joy, anger, sadness, etc.

  Example: To interpret user emotions, for example, social media posts may be subjected to emotional tone recognition.

- Intent Analysis:
  Description: Determining the text's intended meaning, be it neutral, positive, or negative.

  Example: Analysing customer support requests to understand if the user's intent is to seek help, express dissatisfaction, or give positive feedback.

- Irony and Sarcasm Detection:
  Description: Knowing when a person is being ironic or sarcastic when their sentiment is contradictory to what they are saying irony or sarcasm. You can use both irony and sarcasm in your speech. A figure of speech called irony means the exact opposite of what is said. There is a type of irony called sarcasm that is meant to criticize someone.

  Example: Recognizing ironic comments made on social media.

- Comparative Sentiment Analysis:
  Description: Comparing sentiments between topics, products, and entities.

  Example: Using social media discussions to compare public sentiment towards two competing brands.

- Temporal Sentiment Analysis [1]:
  Description: Analysing how sentiments change over time, identifying trends and shifts. Temporal Sentiment Analysis is a valuable tool for summarizing events by considering both sentiment and time. Causal relationships serve as valuable tools for both identifying the cause and effect of events and predicting their outcomes. When these two ideas are combined, the resulting event prediction model is more accurate in predicting the emotion and time between upcoming occurrences.

  Example: Monitoring how people feel about a politician on social media during an election campaign.

- Domain-Specific Sentiment Analysis:
  Description: Customizing Sentiment Analysis models for specific industries or domains.

Example: Developing Sentiment Analysis tailored for healthcare to understand patient sentiments in medical records.

- Fine-Grained Sentiment Analysis:
  Description: Providing more nuanced sentiment classifications, considering a broader range of sentiments beyond just positive, negative, or neutral.
  Example: Distinguishing between sentiments like excitement, disappointment, satisfaction, etc., in product reviews.

These types of Sentiment Analysis cater to different needs and applications, allowing organizations to gain deeper insights into the sentiments expressed in textual data.

We can perform sentiment analysis task at different levels. Figure 1.2 shows the relationship between different levels of sentiment analysis.

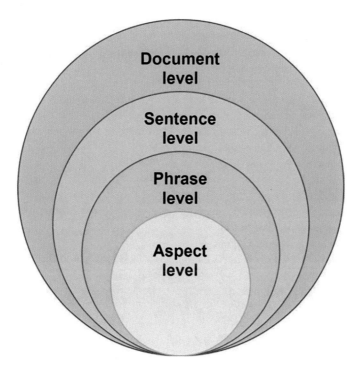

Fig. 1.2 Levels of Sentiment Analysis

1.3 Role of Sentiment Analysis [2, 4, 13]

Sentiment Analysis is a critical process that aims to recognize and interpret the emotions, opinions, and attitudes expressed within textual data. Sentiment Analysis plays various roles across different domains and industries, offering valuable insights into human opinions and emotions expressed in textual data.

Some significant roles of Sentiment Analysis are as follows:

- Customer Feedback Analysis:
  Task: Analysing customer reviews, comments, and feedback to understand the sentiment towards products or services.
  Importance: Helps businesses improve products, address concerns, and enhance customer satisfaction.
- Brand Monitoring:
  Task: Monitoring social media and online mentions to gauge the sentiment associated with a brand or company.
  Importance: Allows businesses to manage their online reputation, identify trends, and respond to customer concerns.
- Market Research:
  Task: Analysing sentiments in market-related discussions to understand consumer preferences and trends.
  Importance: Aids businesses in making informed decisions, launching targeted campaigns, and staying competitive.
- Political Analysis:
  Task: Evaluating sentiments expressed in political speeches, social media, and news to understand public opinion.
  Importance: Assists political campaigns in gauging public sentiment, identifying key issues, and adapting strategies.
- Customer Service Optimization:
  Task: Analysing customer support interactions to identify sentiments and improve service quality.
  Importance: Helps organizations address customer concerns promptly, enhancing overall customer experience.
- Product Development:
  Task: Assessing sentiments in user feedback and surveys to guide product improvements and innovations.
  Importance: Facilitates the development of products that align with customer preferences and needs.
- Employee Engagement:
  Task: Analysing sentiments in employee feedback and communication to understand workplace satisfaction.
  Importance: Supports human resources in creating a positive work environment and addressing employee concerns.
- Financial Market Analysis:

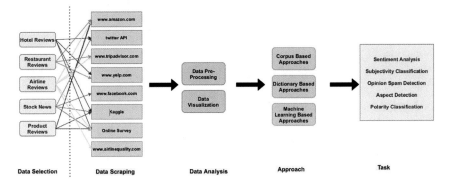

Fig. 1.3 Process of Sentiment Analysis

Task: Analysing sentiments in financial news, reports, and social media to understand market sentiment.
Importance: Helps investors make informed decisions and understand market dynamics.

- Healthcare:
Task: Analysing sentiments in patient feedback, reviews, and medical records to understand patient experiences.
Importance: Aids healthcare providers in improving patient care and addressing concerns.

- Educational Data Mining:
Task: Analysing sentiments in educational content and learner feedback to enhance the learning experience.
Importance: Supports educators in adapting teaching methods and improving overall educational outcomes.

Figure 1.3 shows the process of Sentiment Analysis [7, 13].

These diverse roles demonstrate the versatility and applicability of Sentiment Analysis in extracting meaningful insights from textual data across various sectors. By understanding sentiments, organizations can make data-driven decisions, improve communication, and enhance overall performance.

1.4 Educational Data Mining (EDM) [14]

Educational Data Mining (EDM) is the most emerging academic and research field of computer science which extracts, explores, and exploits data to infer valuable information in the educational context by using data mining (DM) techniques [15]. It combines artificial intelligence techniques to implement data mining, machine learning, and statistical analysis.

Table 1.1 EDM with different types of stakeholders

Users	Objectives for using Educational Data Mining
Learners/learners	Personalized e-learning and recommendation
Educators/teachers/instructors/tutors	Feedback from learners and management
Course developers/educational researchers	Evaluation and maintenance of courseware
Organizations/learning providers/universities/ private training companies	Enhancing decision-making in higher learning institutions
Administrators/network or system administrators	Carving the system optimally to use and organize available institutional resources and their educational offers

It has generated many new ideas related to measuring of learner's needs, learning and teaching pedagogies, arranging learning materials and its packaging, finding and selecting the best available learning materials for efficient and effective planning and implementation of educational processes [16]. Table 1.1 shows the specific purposes related to different stakeholders in the education system.

Figure 1.4 shows educational activities of all the stakeholders, Educational Data Mining techniques used in those activities and analysis of the patterns.

1.5   Technology Enhancement in Educational Data Mining [17, 18]

As of now the educational institutions and their policies are being rapidly adapted according to the trends based on the dynamic changes in the world of education system like print to digital, choices of learning skills and methodologies, changes in learning materials and related products, etc. [19]. They are coping with the changes in the learning methods, i.e. shifting from teacher to learner centric, physical classroom to digital platforms, one to one or one to many through digital gadgets via internet and recorded teaching or online real time basis by using different multimedia tools. Point-to-point delivery of learning is the smart way of learning in a very cost-effective manner. It improves learning experience, learning performance, and overall learning satisfaction. It facilitates a fast adaptive, i.e. learning-on-fly mechanism and indeed, very instrumental in leveraging the use and scope of technology ensuring time and cost-effectiveness. The teacher or system-instructor identifies learning needs, prepares and provides customized learning materials dynamically to the maximus satisfaction of the learners.

The technological and digital disruptions led by internet booms and telecom innovations have augmented the transmission of skills, information and knowledge among learners and educators beyond boundaries through the network of computers supported by internet/intranet, etc. [20]. E-learning is an education system based on the internet or intranet for restricted learning within an organization in which

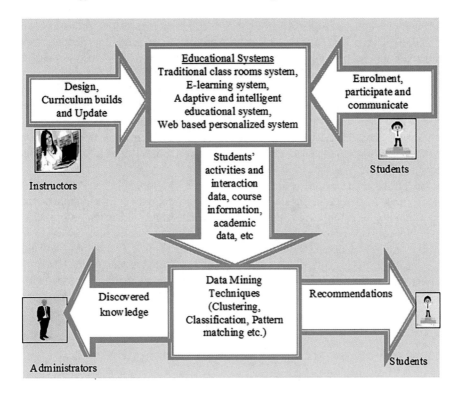

Fig. 1.4 Cycle of applications of Educational Data Mining in education system

e-teaching, in the form of asynchronous, synchronous and hybrid system; imparts education to individual or masses in different ways and especially adaptive now a days. A large amount of data and huge amounts of information are readily made available by the system as and when needed to different stakeholders through various online channels in e-education system. It is the smooth and continuous process leading to better teaching–learning interactions in the way desired to achieve the maximum benefits. The faster availability of voluminous information is a boon for the learners and educations to plan and design their learnings in the most preferable ways. On the other hand, it may be equally damaging if the streams of information are not channelized properly. As we know, unstructured and uncontrolled information may cause interruptions in the learning process, hinder the interest of stakeholders, and badly affect the entire educational system.

However, some data mining techniques address the learning problems suitably [21] the details are as follows: Classification of Learners based on their profile data, qualities inherited, preferences opted, their past performances, etc. By analysing these data, some more advanced findings can be detected like irregular learning patterns revealing their behaviours based on selection/interaction/navigation of study materials, identifying their cluster and adaptable approaches based on usages of the system for the optimization of e-learning needs and capacities. Hence, the assessment of learners can be easily found based on their activities through learning processes and using data mining techniques. The assessment derived from such closer evaluation mechanisms superimposes the traditional education system with a big margin.

The widespread uses and acceptability of internet-based education has opened the floodgate of distance education and is playing a pivotal role in the mainstream education system. It is popularized because many new e-learning models, processes and systems have been developed. The new findings and their implementations have enabled e-learning to be a successful pursuit and established it as a future learning model.

The efforts made by e-learning researchers and developers in EDM have facilitated the new learners and educators to adopt new technologies and their applications. The technologies capture and generate phenomenal strings of data and information. They are constantly helping to improve the e-learning system and processes by deciphering knowledge patterns extracted from Data Mining Techniques into valuable information. Therefore, the main objective is to derive valuable information to enhance learning experience for the learners and upgrade their knowledge/skills/ attitudes based on the patterns of their system usage and behavioural analysis.

1.6 Applications of Educational Data Mining EDM [22, 23]

- Analysing data patterns and recapitalizing further
- Aggregating suggestions, information and feedback and verifying them for better orientation of educators or system instructors
- Advising learners to improvise their learning approaches adaptively
- Analysing and predicting learner's performance or grades or success or failures or promotions
- Creating learning models for individual learner or group of learners providing same learning materials or adaptive point-to-point learning
- Finding and analysing undesirable learner's behaviours from internal and external sources like social media platforms
- Identifying, introducing, and implementing cognitive maps or designing related courseware
- Assessing, structuring, and scheduling the learning processes in different educational system like traditional, web-based, intelligent tutoring and adaptive e-learning system
- Formulating and advising recommender system based on EDM for learners to take better decisions by selecting the best courses/facilities from the pool of available options [24].

Table 1.2 Comparison between business and educational domain

Business domain	Educational domain
Quest for most profitable customers	Quest for learners consuming most credit hours
Quest for the most frequent customers	Quest for ones who return for more classes
Quest for loyal customers	Quest for consistent learner
Quest for customers likely to purchase	Quest for alumnus likely to donate/pledge more
Quest for the customers likely to defect to my rivals	Quest for courses that attract more learners

1.7 Data Mining in Business and Educational Domain [25]

We know that data mining techniques are used to take better decisions in different domains. However, it is more popular in education and used as a tool of business intelligence.

Table 1.2 shows users and their types along with the information required in business and educational domain.

1.8 Terminology Used in Educational Data Mining [26]

Some terms and terminology are used in this chapter related to Educational Data Mining for better understanding. The most frequently used words are as follows:

- **Learning Gaps**: A learning gap is a default or discrepancy or anomaly or difference between what was expected and what is found [27]. So, **the learning gap** is a scale to find out the differences and remedies to remove obstacles for maximizing effective learning and enhance learning experiences, i.e. the way the brain learns automatically, and the way learners are being taught in the classroom.

  - **Types of Learning Gaps**:
    i. **Individual**: Personal development program is to develop required level of knowledge, skills, attitudes, aptitudes, and preferences to mitigate learning gap.
    ii. **Operational**: Operational learning gaps can be reduced by fulfilling gaps found in terms of objectives, shared vision, team cohesiveness, common set of core values, etc.
    iii. **Organizational**: Organizational learning gaps can be mitigated by doing better strategic planning, review and control of performance indicators, etc.

       - **Individual Learning Gaps**: After starting the course, learners are in the state of indecisiveness about the course which is the most suitable for them and what are the other alternative options for them. Even after selecting, they may continue that and complete the course or may leave in between and start another course. It is very difficult to find a suitable course without any recommender system. Even an excellent teacher does not guarantee learner's success across the courses and educational systems. It depends upon learner's individualities, traits, characteristics, and interaction with learning objects. Individual gaps of the individual may be addressed suitably by improving on following factors:

- **Knowledge Gaps**: can be detected and upgraded through exams, tests, quizzes, and assignments. In digital media, computers and consoles are highly cognitive and packed with very powerful tools to detect gaps and advise suitable solutions.
- **Skills Gaps** are detected by identifying core competencies, i.e. workplace efficiency, etc. and improved by Exercise, Lab Simulation, Case Study, Applications, etc.
- **Aptitude and Preferences Gaps**: can be detected by quantitative and observable evidence, measured through the trainings/sessions based on behavioural science and soft skill development (BS Test).

- **Learning Style Theory**:
  There are many learning theories supporting different styles of learning. They describe that the learners learn differently and attain required knowledge and skill sets for working professionally in better way. They learn in different ways because of their different preferences of learning styles. There are different methodologies and classification studies on LS. One of LS's classifications have been proposed by Felder [28], Kolb [29] and many other scholars [28]. Most of the researchers classified them into characteristic groups and sub-groups based on certain assumptions, trends, patterns, and methodologies.
- **Learning Process**:
  The learning preferences of learners depend on the distinct nature of learners, different mode of thinking, and different abilities in co-relating and creating things, also discussed as factors responsible for individuality [30]. It is always challenging for the tutors to modify and adapt teaching strategies, materials, methodologies, and processes. This is also encountered by the computer system in the same way. The learning process improves dramatically when the teaching is being imparted in tune with the preference of the learner's style of learning. The outcomes/outputs depend upon the input variables and preferred way of testing knowledge, skill sets, and applications accordingly. The input or output may vary and suitability with the learner to be closely examined to analyse the information. Finally, we may get the desired co-relation of the individual and their learning style preferences.
- **Learning Behaviour**:
  Learners change their behaviour over a period during the learning processes. The changes occur on account of changes in ever changing psychological-emotional state of the mind of the learners while interacting with the e-learning

system. Sometimes, they behave deliberately to act like working against the system, whereas sometimes it may happen by chance or on account of natural stimuli to the triggers put into the system to detect hidden their behavioural aspects. Boredom, frustration, motivation, concentration, tiredness, etc. are prime emotional quotients guiding the learning journey and helping system and tutors to identify important behavioural aspects which further enabled with to make or mark the effectiveness of the learning process [31].

Identifying and tracking exhibited Learning Behaviours in real-time scenario is a very critical and important job. Authors Conati and Gutl et al. in [32] proposed theories to modelling learner's influence and structured the concepts to evaluate various emotional phases during the interactions and their cause-effect analysis. They applied special sensing scanners to detect changes in behaviour (following the Ortony, Clore, and Collins cognitive theory of emotions). In [32] other study, researchers presented a Dynamic Fuzzy Petri Net inference engine tool that observed "browsing time" and "browsing count" of learners' interactional exchange within the system and in related environment. Additionally, they noted that the amount of time a learner spent on a particular learning object indicated whether the learner was either completely engaged and involved in it and completed the task in no time, if it required a great deal of time but was a very interesting subject for them, or if they were extremely puzzled by it and unable to decide whether to continue reading or quit it in the middle. Many of the authors observed that the learners' motivation plays a vital role in ensuring learning efficiency in many learning theories. It dominates the least preferable styles more visibly. It is interlinked to emotionally charged, self-inclined, focused, high level of concentration, worried, interactively by giving feedback, system generated recommendation, etc. A technique used in "Discovering learner preferences in e-learning" by Carmona et al. [33] is interesting mechanism to study to implement to find out original learner's preferences. Since adaptivity eases out learning process, smoothen the entire learning journeys and enable learners to learn fast and complete the task in less time with better scores. Better performances again motivate them to choose the same path and this path may become their preferable way of learning. These comprehensive ideas are applied to understand the leaner's real problems for overall development of the educational system.

1.9 Sentiment Analysis in Educational Data Mining

1.9.1 Sentiment Analysis and Decision-Making [34–36]

The Sentiment Analysis of different stakeholders, including parents, learners, and educators, provides crucial insights into educational feelings, views, and attitudes, so contributing significantly to educational decision-making.

The following are significant characteristics of the function that Sentiment Analysis plays in the process of making educational decisions.

1.9.1.1 Learner Feedback and Engagement [34, 37]

The analysis of learner feedback, whether provided by forums, surveys, or other means, is assisted by Sentiment Analysis. It exposes the opinions of learners about the overall learning experiences, course materials, and education quality.

Decision Support: Based on the outcomes of Sentiment Analysis, decision-makers are empowered to make smart decisions on modifications to the curriculum, teaching methods, and overall strategies aimed at strengthening learner engagement.

1.9.1.2 Teacher Evaluation and Support [38]

- Performance Assessment: By analysing feedback from learners and colleagues, Sentiment Analysis may be utilized to evaluate the performance of educators. Positive sentiments may serve as indicators of successful teaching methodologies, while negative sentiment may indicate inadequacies that require that need to be improved.
- Professional Development: Professional development programmes can be informed by the identification of areas in which instructors could benefit from further training or support using Sentiment Analysis.

1.9.1.3 Institutional Reputation Management [38]

- Monitoring Public Perception: For monitoring the public's opinion of academic institutions, Sentiment Analysis is useful. This is the analysis of sentiments expressed in news articles, social media posts, and other online platforms to gain a more comprehensive understanding of the public's perception of the organization. Utilizing the outcomes of sentiment research, institutions may formulate strategic decisions aimed at strengthening their reputation, addressing complaints, and improving communication.

1.9.1.4 Admission Processes [39]

- Sentiment Analysis for College Applications: Sentiment Analysis can be used to analyse sentiments conveyed in college application essays, recommendation letters, and interviews. This can help with the college admissions decision-making process.

- Identifying Issues: To enhance the entire admissions process, decision-makers can identify and resolve concerns or difficulties that prospective learners have brought to light.

1.9.2 Crisis Management

Early warning can be provided by Sentiment Analysis through the identification of negative emotions or issues stated by teachers or learners. This capability allows decision-makers to promptly respond to challenging risks, so preventing their progression into crucial situations.

- Proactive Decision-Making: By using Sentiment Analysis, institutions may take action to address emerging issues and reduce the risk for reputational harm.

1.9.3 Personalized Learning [40]

An analysis of learner sentiments on their engagements with online learning systems may give valuable information regarding their individual learning preferences and challenges. Decision-makers may use this information to improve instructional strategies and individualize learning experiences.

Enhancing Learner Performance—Institutions can use data-driven decision-making to improve learner performance by changing educational strategies in accordance with the outcomes of Sentiment Analysis.

In summary, Sentiment Analysis within the education sector facilitates well-informed and data-driven decision-making that enhances the educational experience for stakeholders and learners while also permitting for a proactive and adaptable response to challenges and opportunities in the vibrant educational environment.

1.10   Issues and Challenges in Sentiment Analysis [41]

Sentiment Analysis encounters specific challenges and issues, which will be highlighted here and discussed in Chap. 5:

- Ambiguity and Contextual Understanding [42]:
  Due to the ambiguity of language and the need for contextual analysis, accurately classifying sentiment is a challenging task. Words can assume diverse senses based on the context.
- Handling Sarcasm and Irony [43]:
  Identifying sarcasm and irony is a constant challenge for Sentiment Analysis models, as the expressed sentiment may contradict the true meaning of the words.

- Data Imbalance and Bias [41]:
  Imbalances in sentiment-labelled datasets can lead to biased models where specific opinions are excessively represented, leading to erroneous predictions, especially for minority feelings.
- Cross-Domain Adaptation:
  Models trained on sentiment data from a certain domain may exhibit limited generalized capabilities when applied to different domains, leading to a lack of adaptability in various contexts.
- Multilingual Sentiment Analysis [44]:
  Performing Sentiment Analysis in multilingual environments is challenging because of variations in language, cultural distinctions, and a scarcity of labelled data for less frequently used languages.

It is possible for sentiment to change over time, leading to outdated or inaccurate sentiment estimations.

- Ethical Concerns and Privacy:
  The application of Sentiment Analysis in sensitive situations creates ethical concerns around user privacy and the potential exploitation of sentiment results [36].

1.11 Conclusion

This chapter explores an overview on the fundamental concepts of Sentiment Analysis, describing its relevance and effectiveness. By identifying and summarizing insights from both Educational Data Mining and Sentiment Analysis, this chapter makes a scholarly and thought-provoking contribution to unwind a broader discussion on enhancing educational experiences through data-driven approaches, enabling the possibility of adaptive decision-making and generating improvised learning outcomes finally.

References

1. Preethi PG, Uma V, Kumar A (2015) Temporal sentiment analysis and causal rules extraction from tweets for event prediction. Procedia Comput Sci 84–89. https://doi.org/10.1016/j.procs.2015.04.154
2. Zhou J, Ye JM (2023) Sentiment analysis in education research: a review of journal publications. Interact Learn Environ. https://doi.org/10.1080/10494820.2020.1826985
3. Evans D (2002) Emotion: the science of sentiment. Am J Orthopsychiatry 72(4). https://doi.org/10.1037//0002-9432.72.4.601
4. Altrabsheh N, Gaber MM, Cocea M (2013) SA-E: sentiment analysis for education. Front Artif Intell Appl 255:353–362. https://doi.org/10.3233/978-1-61499-264-6-353

5. Montero CS, Suhonen J (2014) Emotion analysis meets learning analytics—online learner profiling beyond numerical data. In: ACM international conference proceeding series, vol 2014, Nov 2014, pp 165–169. https://doi.org/10.1145/2674683.2674699

6. Wiebe JM (1994) Tracking point of view in narrative. Comput Linguist 20(2):233–287. Accessed 26 Jan 2024. [Online]. Available: https://aclanthology.org/J94-2004

7. Wankhade M, Rao A, Kulkarni C (2022) A survey on sentiment analysis methods, applications, and challenges. Artif Intell Rev. https://doi.org/10.1007/s10462-022-10144-1

8. Crystal D, Quirk R (2021) Systems of prosodic and paralinguistic features in English. https://doi.org/10.1515/9783112414989

9. Pang B, Lee L (2008) Opinion mining and sentiment analysis. Found Trends Inf Retr 2(1–2):1–135. https://doi.org/10.1561/1500000011

10. Sindhu I, Muhammad Daudpota S, Badar K, Bakhtyar M, Baber J, Nurunnabi M (2019) Aspect-based opinion mining on student's feedback for faculty teaching performance evaluation. IEEE Access 7:108729–108741. https://doi.org/10.1109/ACCESS.2019.2928872

11. Hajrizi R, Nuçi KP (2020) Aspect-based sentiment analysis in education domain. [Online]. Available: http://arxiv.org/abs/2010.01429

12. Hovy EH (2015) What are sentiment, affect, and emotion? Applying the methodology of Michael Zock to sentiment analysis, pp 13–24. https://doi.org/10.1007/978-3-319-08043-7_2

13. Zhang W, Li X, Deng Y, Bing L, Lam W (2023) A survey on aspect-based sentiment analysis: tasks, methods, and challenges. IEEE Trans Knowl Data Eng 35(11):11019–11038. https://doi.org/10.1109/TKDE.2022.3230975

14. Sweta S (2021) Educational data mining techniques with modern approach. In: Modern approach to educational data mining and its applications. Springer, pp 25–38. https://doi.org/10.1007/978-981-33-4681-9_3

15. Peña-Ayala A (2014) Educational data mining: a survey and a data mining-based analysis of recent works. Expert Syst Appl 41(4 PART 1):1432–1462. https://doi.org/10.1016/j.eswa.2013.08.042

16. Baker RSJD, Yacef K (2009) The state of educational data mining in 2009: a review and future visions. J Educ Data Min 1(1):3–17

17. Sweta S, Lal K (2017) Personalized adaptive learner model in E-learning system using FCM and fuzzy inference system. Int J Fuzzy Syst 19(4):1249–1260. https://doi.org/10.1007/S40815-017-0309-Y

18. Sweta S, Lal K (2015) Web usages mining in automatic detection of learning style in personalized e-learning system. Adv Intell Syst Comput 415:353–363. https://doi.org/10.1007/978-3-319-27212-2_27

19. Sweta S, Lal K (2014) Adaptive e-learning system: a state of art. Int J Comput Appl 107(7):13–15

20. Romero C, Ventura S (2010) Educational data mining: a review of the state of the art. IEEE Trans Syst Man Cybern Part C (Appl Rev) 40(6):601–618. https://doi.org/10.1109/TSMCC.2010.2053532

21. Jindal R, Borah MD (2013) A survey on educational data mining and research trends. Int J Database Manag Syst (IJDMS) 5(3):53–73. https://doi.org/10.5121/ijdms.2013.5304

22. Romero C, Ventura S (2013) Data mining in education. Wiley Interdiscip Rev Data Min Knowl Discov 3(1):12–27. https://doi.org/10.1002/widm.1075

23. Özyurt Ö, Özyurt H (2015) Learning style based individualized adaptive e-learning environments: content analysis of the articles published from 2005 to 2014. Comput Human Behav 52:349–358. https://doi.org/10.1016/j.chb.2015.06.020

24. Zapata A, Menéndez VH, Prieto ME, Romero C (2015) Evaluation and selection of group recommendation strategies for collaborative searching of learning objects. Int J Hum Comput Stud 76:22–39. https://doi.org/10.1016/j.ijhcs.2014.12.002

25. Sweta S (2021) Educational data mining techniques with modern approach. In: Springer briefs in applied sciences and technology, pp 25–38. https://doi.org/10.1007/978-981-33-4681-9_3

26. Sweta S (2021) Modern approach to educational data mining and its applications. Accessed 29 Jan 2024. [Online]. Available: https://doi.org/10.1007/978-981-33-4681-9.pdf

27. Lucke U, Rensing C (2013) A survey on pervasive education. Pervasive Mob Comput 14:3–16. https://doi.org/10.1016/j.pmcj.2013.12.001
28. Felder R, Silverman L (1988) Learning and teaching styles in engineering education. Eng Educ 78:674–681. https://doi.org/10.1109/FIE.2008.4720326
29. Kolb DA (1981) Learning styles and disciplinary differences. In: Responding to the new realities of diverse students and a changing society, pp 232–255. https://doi.org/10.1016/S0002-8223(97)00469-0
30. Sweta S (2015) Adaptive and personalized intelligent learning interface (APIE-LMS) in e-learning system. Int J Appl Eng Res 10(21):42488–42492
31. Truong HM (2015) Integrating learning styles and adaptive e-learning system: current developments, problems and opportunities. Comput Human Behav. https://doi.org/10.1016/j.chb.2015.02.014
32. Botsios S, Georgiou D (2008) Recent adaptive e-learning contributions towards a "standard ready" architecture. e-Learning
33. Carmona C, Castillo G, Millán E (2007) Discovering student preferences in e-learning. CEUR Workshop Proc 305:33–42
34. Ravi M, Johnson SJ. Analysis of student feedback on faculty teaching using sentiment analysis and NLP techniques
35. Mejova Y (2009) Sentiment analysis: an overview comprehensive exam paper. [Online]. Available: http://www.pewinternet.org/Reports/2009/15-The-Internet-and-Civic-Engagement.aspx
36. Zhou J, Ye JM (2020) Sentiment analysis in education research: a review of journal publications. Interact Learn Environ. https://doi.org/10.1080/10494820.2020.1826985
37. Liu NF, Carless D (2006) Peer feedback: the learning element of peer assessment. Teach High Educ 11(3):279–290. https://doi.org/10.1080/13562510600680582
38. Romero C, Ventura S (2020) Educational data mining and learning analytics: an updated survey. Wiley Interdiscip Rev Data Min Knowl Discov 10(3). https://doi.org/10.1002/widm.1355
39. Ahmed LM, Hussein GS, Zaied ANH (2020) A survey on sentiment analysis algorithms and techniques for Arabic textual data. Fusion Pract Appl. https://doi.org/10.54216/FPA.020205
40. Medhat W, Hassan A, Korashy H (2014) Sentiment analysis algorithms and applications: a survey. Ain Shams Eng J. Accessed 26 Jan 2024. [Online]. Available: https://www.sciencedirect.com/science/article/pii/S2090447914000550
41. Liang W et al (2022) Advances, challenges and opportunities in creating data for trustworthy AI. Nat Mach Intell 4(8):669–677. https://doi.org/10.1038/s42256-022-00516-1
42. Breitung C, Kruthof G, Müller S (2023) Contextualized sentiment analysis using large language models. SSRN Electron J. https://doi.org/10.2139/SSRN.4615038
43. Irony vs. sarcasm: types and differences | YourDictionary. Accessed 26 Jan 2024. [Online]. Available: https://www.yourdictionary.com/articles/irony-sarcasm-difference
44. Dashtipour K et al (2016) Multilingual sentiment analysis: state of the art and independent comparison of techniques. Cognit Comput 8(4):757–771. https://doi.org/10.1007/S12559-016-9415-7/TABLES/2

Chapter 2
Application of Sentiment Analysis in Diverse Domains

2.1 Introduction

Sentiment Analysis (SA), which also referred as opinion mining, has ability to extract valuable information from text data. Because of this powerful feature, it has been applied in many fields. This chapter pertains usefulness of the Sentiment Analysis in the various domains. The profound success of learning processes after integrating and collaborating Sentiment Analysis have open wide spectrum to explore and exploit the accuracy and efficiency parameters in diverse domains. From social media and customer reviews to healthcare and finance, the scope of Sentiment Analysis extends across diverse sectors, reshaping the way we interpret and respond to human emotions.

Here are some prominent fields where Sentiment Analysis is applied.

2.2 Sentiment Analysis in Different Domains [1]

2.2.1 Business and Marketing

Applying Sentiment Analysis in the disciplines of business and marketing offers useful insights into customer preferences, opinions, and overall sentiments towards products, services, or the brand. Businesses employ Sentiment Analysis to assess customer sentiments regarding their products or services. This facilitates brand management, enhances marketing plan optimization, and prepare for understanding of customer preferences.

Examples: Social Media Monitoring is helpful to gain insights on customer sentiment regarding launching a new product.

© The Author(s), under exclusive license to Springer Nature Singapore Pte Ltd. 2024
S. Sweta, *Sentiment Analysis and its Application in Educational Data Mining*,
SpringerBriefs in Computational Intelligence,
https://doi.org/10.1007/978-981-97-2474-1_2

2.2.1.1 Methods Used in Business and Marketing [2, 3]

- Social Media Monitoring:

  - Approach: Apply Sentiment Analysis tools to track social media platforms, blogs and forums, for reference to your brand, products, or industry-specific terms.
  - Example: Gain real-time insights into customer perceptions of your business, detect developing trends, and interact with your audience based on their opinions.

- Customer Reviews Analysis:

  - Approach: Apply Sentiment Analysis on reviews of customer mentioned on e-commerce platforms, review sites, or your website.
  - Example: Get valuable knowledge of the specific features or qualities of products or services that clients perceive or find challenging. Identify possible opportunities for enhancement and monitor the effects of improvements over a specific period.

- Brand Monitoring:

  - Approach: Continuously observe articles, blogs, and social media platforms to identify any mentions or discussions related to your brand.
  - Example: Assess the overall reputation of your brand, identify possible public relations crises, and understand the effectiveness of marketing activities in connecting with your target audience.

- Analysis of Product Launch:

  - Approach: Conduct Sentiment Analysis on social media discussions and reviews to evaluate the emotional tone before, during, and after the launch of a product.
  - Example: Assess and evaluate the public opinions and their expectations related to the initial response and performance of the products and services and identify the underlying new opportunities after post-launch analysis and further modifications in view of contemporary marketing trends.

- Competitors Analysis:

  - Approach: Exploit the utility of Sentiment Analysis to assess, analyse, and evaluate references those compete with the companies or different competitors on different digital platforms like social media, online review systems etc.
  - Example: Understand the inside stories and new insights about the perception of the brand value in comparison to the competitors based on study of customer's sentiment. It can be easily evaluated the strengths and weaknesses of the different competitors.

- Evaluation of the Campaign:

- Approach: Analyse and evaluating sentiments of marketing team while making marketing efforts by finding the moments of the truths in social media contents and customer responses across all available platforms like website or different assessment engines.
- Example: Assess, evaluate, and take corrective steps which will enhance efficiency of marketing efforts and help to develop new skill sets to intensify marketing efforts with new ideas, understand customer's need and accordingly change responses to make the interactions successful in terms of conversion of leads into revenue generating business propositions. It also helps to change the goalposts in midway and do adaptive corrective measures to get the desired outcome more precisely and satisfying for both. It really helps to get win–win resulted outcomes.

2.2.1.2 Advantages of Sentiment Analysis (SA) in Business and Marketing [1]

- Insights About Consumer's Behaviour: Leverage the complete understanding of opinions, preferences, and expectations of consumers and all stakeholders. It enables focussed customer-centric decision-making processes.
- Managing Reputations: The proactive initiatives to effectively manage and protect the brand's online/offline reputation by mitigating negative opinions and promoting positive ones across the board.
- Product Improvement: Customer's satisfaction is increased by implementing product line improvement based on analysis of their primary inputs to identify specific areas for improvement in a time-bound manner.
- Competitive Advantage: By understanding the emotional queues from the customers about the competitive edge of the product or services and how our brand compares and competes among the competitors' value propositions.
- Real-Time Decision Making: By real-time flexible decision-making to address new information, upcoming trends, challenges, threats, or opportunities; better results can be facilitated.
- Campaign Optimization: Ensure the effective executions of the marketing efforts and strategies in short term as well as in long term by using real-time Sentiment Analysis of the consumers to align with the corporate goals objectives.
- Customer Engagement: Enhance customer engagement and customer experience by promptly responding to the feedback given by customers and exhibiting delightful communication to the extent of customer happiness.
- Risk Mitigation: Analyse, detect, and promptly manage all possible public related critical issues by monitoring the customer's attitudes to minimize the probability of harm/loss to the organizational reputation.

In short, we can say that the Sentiment Analysis provides a data-driven mechanism for detecting, comprehending, and addressing customer's sentiments, perceptions, and other emotional traits to improvise the brand value and organizational efficiencies and achievements while taking business and marketing efforts. Importantly, how the course corrections would be done to add value in the entire process.

2.2.2 Customer Feedback and Reviews [3]

Integrating the Sentiment Analysis in the domain of customer's feedback, perceptions and reviews is a wonderful strategic approach to generate most important insights from the huge unstructured data produced by the consumers. Sentiment Analysis is widely used for the assessment and identification of customer feedback during real-time interactions and its evaluations to address negative sentiments. It provides vital information and relevant data of the company regarding customer's satisfaction and remaining areas which require further improvement.

Example: Assessing internet feedback and evaluations of a restaurant to understand and identify customer sentiments regarding the ambiance, cuisine, delicacy, cleanliness, service, and pleasant environment.

2.2.2.1 Process of Improvement Customer Feedback and Reviews

- Data Collection Process: Obtaining customers feedback, compiling data and evaluating from diverse sources, such as online review on different websites of corporate or business aggregators through different apps and platforms, any interactive data available on social media, questionnaires asked to customers by online system or AI generated software and interactive tools linked to customer support interactions. Thorough and comprehensive data collection and compilation guarantees a wide-ranging representation of customers' opinions which give consolidated insights and help to modulate emotional quotients to gain maximum.
- Text Preprocessing: Implementing data cleaning tools and preprocessing on the text, audio and video files by eliminating noise like extraneous characters, irrelevant data and unavoidable special characters. It improves the accuracy and quality of Sentiment Analysis algorithms and hence guarantees trustworthy results.
- The Sentiment Analysis Algorithm involves the utilization and implementation of Sentiment Analysis derived from the interactive data or tool to analyse the preprocessed text/audio/video and ascertain the changes in the sentiments, which may be found as positive, negative, or neutral. The result can further be fine-tuned based on new adaptive inputs. Exhibits the automatic transmission of sentiment classifications, allowing them for using effectively and efficiently for extensive amounts of customer's feedback and understanding sentiments of the group of customers and their preferences.
- The Techniques of Aspect-Based Sentiment Analysis: To get more comprehensive and in-depth insights, aspect-based Sentiment Analysis can be implemented to detect human sentiments linked to the specific elements, defining ideas or differentiating characteristics of the product or service. Conceptualize the specific areas of

excellence of Sentiment Analysis and address critical deficiencies in products or services, enabling focused and time-bound development in products or services.

- Visualizations and Dashboards Process: The process involves creating visual manifestations and interactive displays that provide a concise and fact-finding overview of how the analysis of sentiment patterns have emerged over the time and across the different characteristics. Offers a concrete and easily comprehensible overview of customers' feelings, facilitating critical decision-making and the further development of business and marketing strategies.
- The Categorization of Feedback: This system involves classifying various inputs into distinctive subjects, scenarios, or themes; enabling a more detailed comprehensive customers' attitudes related to the parameters mostly affecting the decision-making process. Provides a more targeted and exclusive strategy for resolving customers' concerns, problems, and complaints for enhancing the face value and key facets of the organization.
- Benchmarking Against Competitors: This methodology involves comparing the Sentiment Analysis with the sentiment of other competitors to acquire valuable inputs, patterns and insights into how the brand is received by users. Evaluate the consumer sentiment to identify the competitive edge overs the others to enhance knowledge and desired gain.

2.2.2.2 Advantages of Sentiment Analysis in Customer's Feedback and Reviews

- Customer's Satisfaction and Developed Insights: Gain an in-depth understanding of customer's satisfaction matrix and its influential factors those impacting opinions and generating positive or negative outcomes.
- Value Enhancement of Product/Services: Finding the scope of improvement through consumer's inputs which help in offerings to customers with improved products or services.
- Management of Brand Value and Its Perception: Initiate proactive steps and calculated actions to control and drive the way we need to establish the brand. The Brand's perceptions can be managed by resolving all negative feelings and enhancing favourable ones.
- Customer Engagement: Fostering a favourable relationship by engaging customers in contributory roles and putting proper reward mechanism for encouragements and keeping their interests by promptly responding to their queries, feedback and providing feasible solutions in a meaningful manner as desired.
- Optimization of Marketing Strategy: Optimizing the marketing strategy based on analysis of consumer attitudes and running the marketing campaigns in line with the customer's expectations and best to their satisfactions.
- Competitive Analysis: Acquiring insights about the brand and its associated consumer's sentiment and comparing them to that of the competitors by identifying opportunities to establish a competitive edge.

- Risk Mitigation: Measuring, evaluating, and addressing all potential risks or concerns in the processes at the material time by monitoring attitudes, behavioural traits, rapid action, and resolution techniques.
- Enhance Operational Efficiency: Resolving repetitive problems detected by the Sentiment Analysis to ensure enhancement in effectiveness.

By integrating Sentiment Analysis into the practical approach for analysing and evaluating feedback and reviews, the unstructured data can be converted into mindful insights. It helps in promoting continuous improvement and facilitating customer-centric decision-making.

2.2.3 Sentiment Analysis in Social Media Monitoring [4, 5]

Application: Sentiment Analysis is used by organizations to monitor public opinions and emotions on their brand, competitors, or specific topics on social media platforms. Example: Analysing tweets to understand public emotion during a political event or emergency in the community.

2.2.3.1 Methods to Use SA for Social Media Monitoring

- Selecting an Appropriate Sentiment Analysis Tool: Identifying criteria for selecting a Sentiment Analysis tool that is well-suited to our requirements. Look out technologies that offer support for many social media platforms, languages, and provide real time analysis.
- Examples of tools and technologies include IBM Watson Natural Language Understanding, Google Cloud Natural Language API, as well as specialized social media monitoring tools such as Brand watch and Hootsuite.
- Define Your Objectives: Setting objectives involves clearly defining the specific goals your aim to accomplish via the practice of social media monitoring. Are you interested by the prospect of understanding brand emotion, monitoring campaign performance, or identifying upcoming trends?
- Goal Example: Monitoring sentiment during marketing campaigns, tracking the general perception of your business, and identifying potential issues before they get worse.
- Determine Appropriate Social Media Pathways: Analyse the platforms on social media that are most relevant to your business or brand. Examples of platforms which includes Twitter, Facebook, Instagram, LinkedIn, forums, blogs, and other platforms unique to industries.
- Set up Keyword and Hashtag Tracking: Select relevant keywords and hashtags that are associated with your brand, products, or industry.

- Tracking Tool: Utilize monitoring features provided by your selected Sentiment Analysis tool or utilize social media management services to monitor the specified keywords.
- Understand Sentiment Classes: Use Sentiment classes to define the sentiment categories you want to monitor (positive, negative, neutral), or use more granular categories linked with your goals (e.g., enthusiasm, discontent).
- Customization: Certain solutions offer the ability to customize emotion categories according to your individual requirements.
- Participation in Regular Monitoring and Analysis: Maintain a consistent practice of monitoring social media mentions and analysing the sentiment linked with them.
- Real-Time Analysis: Utilize technologies that offer real-time analysis to be informed about current trends and emotions.
- Respond and Engage: Developing a comprehensive engagement plan for effectively and efficiently responding to the sentiments and measuring for further fine tuning the augmented results. Actively interacting with the positive emotions to strengthen the positive encounters and promptly handling the negative emotions to minimize and address all possible problems.
- Customer Engagement: Utilizing all social media platforms as a medium for direct customer connect and long-term relationship-building process.

2.2.3.2   Measuring and Monitoring SA Through Social Media Platforms

- Management of Brand Perception: Getting valuable insight into the online perception measuring processes about the brand and taking proactive actions to effectively control, efficiently showcase, and resultantly improve its reputation and establish better brand value.
- Performance Evaluation: Finding the efficacy of marketing campaigns and strategies to make it successful by deciphering underlying meanings of different keywords and hashtags about the Sentiment Analysis during the ongoing campaigns.
- Competitor Analysis: Identifying, monitoring, and evaluating slow or fast changes and revealing sentiment patterns associated with the competitors, enable to figure out their competitive advantages or disadvantages.
- Product Feedback and Improvement: Receiving valuable customer's opinions, feedback or reviews on the products or services targeted for selling and marketing, allows to improved decisions on basis of the informed data.
- Identify Emerging Trends: Identifying emerging trends, its impacts and taking advantages of all possibilities and alternative options by closely monitoring the attitudes exhibited by the influential or targeted audience.
- Crisis Prevention and Management: Measuring and reducing all potential threats and crises by quickly identifying the negative sentiments and performing preventive actions accordingly to minimize or mitigate loss.

- Customers Engagement and Loyalty: Engaging with customers in the present moment, monitoring value additions, and developing a sense of community by establishing customer's loyalty and continued trustworthiness.
- Data-Driven Decision-Making: Enhancing decision-making process by applying insights gained from the Sentiment Analysis to inform decision makers to make better strategies and take appropriate actions. By incorporating the Sentiment Analysis into the social media platforms and monitoring changes in the led down strategies, the brand management, overall marketing efforts and long-term relationship with customers will improve manifold.

2.2.4 Sentiment Analysis in Finance Markets

- Application: Sentiment Analysis is used in financial markets to analyse news articles, live updates, social media posts and comments, and publications of core financial data in order to understand investor's sentiment and predict market movements accordingly by the analysts and all stakeholders.
- Example: Impact and movement of stock prices can be predicted by analysing trending data and information and news of the stocks in-depth.

2.2.4.1 Methods to Utilizing Sentiment Analysis in Social Media Monitoring

- Data Collection Sources: Gather textual data from several sources including financial news articles, social media platforms (Twitter, StockTwits), financial reports, and online forums. APIs: Utilize APIs offered by financial news sources or social media platforms for accessing real-time data.
- Data Preprocessing: Cleaning Text Data—Remove irrelevant information, punctuation, and standardize text (convert to lowercase). Tokenization is the process of dividing a text into distinct components such as words or phrases. Eliminate Stop Words: Remove often-used terms that don't have significant meaning.
- Sentiment Analysis Techniques: Lexicon-Based Approaches involve the use of sentiment lexicons, which are collections of words accompanied by sentiment scores, to analyse sentiment.
- Machine Learning Models: Analyse the labelled data to train models by applying Machine Learning algorithm such as Naive Bayes and Support Vector Machines for sentiment prediction.
- Deep Learning Models: Utilize complex Deep Learning models, such as LSTM or transformer-based models, to conduct more complex Sentiment Analysis.
- Feature Extraction: Word Embeddings: Convert words into numerical vectors using techniques such as Word2Vec or GloVe.
- Document Embeddings: Documents embeddings are vector representations of entirety documents, such as articles as a vector.

- Sentiment Classification: Utilize thresholding to assign sentiment scores as positive, negative, or neutral.
- Predictions Based on Machine Learning: Sentiments can be classified and predicted based on the applications of Machine Learning models.
- Real-Time Analysis: Measuring and reflecting the Sentiment Analysis on the live or streaming data by capturing the market sentiment as well as changes in real-time.
- API Integration: Using APIs to continuously retrieve, measure, evaluate, and analyse new data and patterns.
- Integration with Trading Algorithms: Implementing the Sentiment Analysis into the algorithmic trading methods to facilitate data-driven trading decisions and its outcomes.
- Market Signals: Incorporating Sentiment Analysis as a key parameter in the trading algorithms.
- Visualization and Monitoring Dashboard: Developing a comprehensive dashboard that displays the visual representations and analysis of the sentiment patterns over a period of time. Alerts: Notifications for any significant changes in the attitude or events as and when occurred.
- Back Testing: Measuring performance by back testing trading strategies based on the historical sentiment related data.
- Optimization: Improving overall strategies by making due adjustments based on the outcomes of back testing and information.
- Risk Management Linked to Association Analysis: Evaluating the association between the sentiment and market movements to determine possible hazards and information to take preventive measures.
- Scenario Analysis: Performing scenario analysis by using various sentiment scenarios recorded over a period of time.
- Combination with Fundamental Analysis or Comprehensive Analysis: Integrating the Sentiment Analysis into the traditional fundamental analysis to find out comparisons to understand market dynamics to take suitable actions.
- Adaptability and Continuous Improvement, i.e. Feedback Loop: Establishing a feedback loop to continuously monitor the improvement of the sentiment predictions by ever evaluating their accuracy and efficacy.
- Model Updates: Regularly updating the Sentiment Analysis models to accommodate the ever-changing market conditions and variable market forces.
- Ethical Issues by Balancing Accuracy and Impact: Striving for the accuracy while taking into account all the ethical issues, particularly in sensitive sectors where spreading disinformation can lead to major consequences and wrong perpetual decision making.

Applying sentiment research in financial markets requires the overall integration of the text processing methodologies, high proficiency in the Machine Learning, and a profound comprehension of the financial markets. By integrating new models, techniques and approaches, the accuracy, acceptability and reliability of the Sentiment

Analysis in the financial decision-making process may be significantly improved as technologies progress and efficiency of the system improvises.

2.2.4.2 Advantages of Sentiment Analysis in the Field of Financial Market

- Market Insight Based on Investor's Opinion: Sentiment Analysis is one of the most valuable mechanisms to assess, analyse, and predict the overall perception of investors, traders, brokers and the public towards the individual financial instrument or in view of the broader market perspectives. Researchers and market analysts can predict market movements and undergoing trends by analysing the attitudes and perceptions conveyed in the articles, on various social media posts and contemporary discussions on financial forums or online apps or live commercial news on TV.
- Risk Management by Early Identification of Risks: Sentiment Analysis enables early identification of potential risks as well as market disruptions through the process of early detection of the changes in the market directly linked to the sentiment. By quantifying risk and incorporating sentiment data, more informative system can be developed to help traders, investors, and brokers to take more informed decisions.
- Algorithmic Trading Tactical Decision: Improving Trading Models—Sentiment Analysis provides an extra layer of knowledge information to the algorithmic trading model, which helps in improving the model. It helps in growth of more advanced and data-based tactical decision system.
- Real-Time Decision-Making Mechanism: Algorithms can adapt running trading decisions in real-time by analysing, evaluating and responding to the ever-changing sentiment patterns.
- Asset Management and Portfolio Optimization: Factoring Sentiment Analysis into the asset management methods, which improvise the optimization of investment portfolios considering the trending market sentiments.
- Identification of Opportunity: It gives ideas to fund managers in identifying opportunities and selecting best alternatives among the pool of assets for investment and growth prospects.
- Automated Analysis of Financial News: It helps in streamlining and evaluating fast moving a large number of financial updates facility wise, sector wise and market wise. Resultantly, it helps in co-ordinated decision making.
- Identification of Market Drivers: Market sentiments are identified based on relevant news items or events influencing market forces. The patterns help in decision making.
- Market Forecasting: Predicting Trends—Sentiment Analysis can improve the ability to predict market trends by detecting consistent trends in sentiments that historically have been linked to specific market movements.
- Volatility Prediction: Helps in predicting potential volatility in the market by comprehending the emotional sentiment of the market participants.

- Investor Sentiment Indices: Creating Indices—Financial institutions can develop sentiment indices that serve as a representation of the total market sentiment, providing consumers a quantitative measure.
- Comparative Analysis: Facilitates the comparison of sentiment indicators with market indices for correlation research.
- Customer Relations for Financial Organizations: Analysing Customer Sentiments—Financial organizations can use Sentiment Analysis to get insights into customer sentiments towards their services, allowing improved customer relationship management.
- Proactive Communication: Facilitates the proactive resolution of consumer complaints raised in public forum.
- Early Warning and Fraud Detection: The Sentiment Analysis can be used to assess, analyse and identify early warning signals as well as detection of the fraud based on patterns revealed from monitoring unusual behaviours, market manipulations and fraudulent behaviours of market makers. It can be detected early on the basis of initial indications and verifying facts based on available historical data.
- Analysis of Macro-Economic Indicators: Sentiment Analysis helps in analysis of trending financial news and how they will affect macro-economic scenario. It provides the insights into the overall economic outlook, future aspects and public sentiments.
- Impact of Policy Changes: The changes in economic or global financial market-related policies and critical decisions taken by regulators or government of different countries affect behavioural aspects of the investors and brokers. Their behavioural patterns give insights to other stakeholders in response to the sentiments of all decision makers.
- Behavioural and Financial Insights: The Sentiment Analysis reveals emotional and behavioural patterns and thus the financial insights gained influences the decision-making by investors and brokers. Their unravelling behavioural patterns help to understand the dynamic market movements to take suitable decisions.
- Market Psychology: Understanding of market psychology is very important for predicting new trends in the market and analysing those predictions after completion of the predicted period.
- Enhanced Decision-Making: Utilizing data-driven decision support through Sentiment Analysis allows for a more objective approach, reducing the requirement for subjective judgments.
- Processing Based on Enhanced Information: Valuable extracts from voluminous unstructured data on account of enhanced information are very useful in effective processing and decision making.
- Adaptation Based on Flow of Latest Financial News: Real-time changes of financial information help to take adaptive decisions which help the traders and investors to quickly analyse the situation and adjust their transactions like buying or selling accordingly in response to dynamic movement of the market-like queues from the breaking news and changing patterns of transactions and overall market perceptions.

When applied wisely, Sentiment Analysis in financial markets improves the analytical capabilities of financial professionals and helps in making more informed data-driven decisions-making processes on facts. However, it is important to take into consideration the limitations and underlying bias associated with Sentiment Analysis techniques and methods.

2.2.5 Sentiment Analysis in Healthcare Sector [6, 7]

The healthcare industry may use the Sentiment Analysis to get valuable information about patient experiences, public opinions of healthcare providers, and sentiment expressed in healthcare-related materials.

- Application: Sentiment Analysis is used in the healthcare industry to analyse patient review and feedback, helping healthcare providers in understanding patient satisfaction and improving services.
- Example: Analysing evaluations of medical facilities helps identify specific areas for improvement in patient care.

2.2.5.1 Methods to Utilizing SA in the Healthcare Sector

- Patient Feedback Analysis: Data Collection—Collect feedback from patients from many sources, such as surveys, online reviews, social media, and patient forums. Preprocessing: It is a method to remove and clean noise and other irreverent data from available text/audio/video/other sources. Sentiment Analysis can be applied to identify patient feedback into some categories like positive, neutral, or negative. Using Trend Analysis identification of repeated feedback good/bad/can't say from the patients indicating some good things or common issues. It helps to get some valuable insights for course correction well in time.
- Social Media Monitoring and Platform Selection: Many healthcare issues or debates occur on social media channels like Twitter, Facebook, or healthcare-specific forums/chat room. With the help of Real-Time Analysis—Real time Sentiment Analysis can be utilized to provide real-time solutions to various stake-holders in terms of healthcare facilities, medical treatments, or addressing public health issues.
- Provider and Facility Evaluation and Review Platforms: Evaluating reviews of emotions expressed on various healthcare websites, forums, chat room, etc.
- Benchmarking: Comparing sentiments associated with various healthcare providers, facilities, machines or professionals with the set benchmarks to identify areas where further improvement, fine-tuning or highlighting can be done to enrich the existing competencies.
- Policy Feedback: Sentiment Analysis can be used to evaluate the feedback, public acceptance and concerns on healthcare policy matters and initiatives.

- Brand Monitoring for Pharmaceutical Company Through Drug Sentiments: Opinion expressed good/bad/other in conversations or indications captured through exhibited sentiments can be analysed to keep track on drugs and pharmaceutical brands about its uses and side effects, if any.
- Enhancing Patient Involvement in Clinical Trials: Sentiment Analysis helps to enhance patient's involvement through expressed views and symptoms during pre- and post-clinical trials.
- Recruitment Sentiments: Analysing the changes in sentiments/feelings associated with the pre and post selection of patients for clinical trials. Using Participant Experiences—Studying how enthusiastically patients are involved in clinical trials, what they stated or what they expressed. It gives valuable insights to enhance effectiveness for all stakeholders.
- Health Communication in the Public Sphere, i.e. Media Analysis: Conducting Sentiment Analysis based on government announcements, experts' opinion, research and development releases, articles from renowned personalities in medical, press releases from pharma companies or other integrators and usual public health announcements to get valuable insights about the public opinions, their concerns and attitudes towards health-related matters.
- Crisis Communication and Emergency Response: Monitoring public sentiments, their genuine demands and sharp reactions during pandemics, health crises or emergencies to have insight into their attitude, supports, concerns, and emotions. Crisis Communication Planning utilizing Sentiment Analysis and its findings to guide how to improve communication tactics throughout the health emergencies and what would be long term plannings.
- Marketing in the Healthcare Industry and Evaluation During Campaigns: Conducting Sentiment Analysis on healthcare products, services and advertising campaigns to evaluate the trending patterns and their influences. Through Brand Perception—getting latest insight from the public's opinions and perceptions about the healthcare brands, products and services through Sentiment Analysis of marketing efforts and initiatives.
- Internal Feedback About Physical and Staff Morale: Internal feedback methods can be used to assess the morale and satisfaction levels of healthcare staff through Sentiment Analysis. Analysing sentiments expressed or associated with workplace wellness programmes and complementary medical facilities provided by employer.
- User Experience in Telehealth Services: Conducting Sentiment Analysis on telehealth services to analyse the patient's experiences and getting further insights. To Improve Opportunities—identifying opportunities for development and improvement in telehealth services by analysing patient's sentiments/feedback expressed or otherwise.
- Diseases Surveillance: Applying Sentiment Analysis upon expressed text/audio/video on social media platforms to swiftly identify early warning signs of disease outbreaks or public health hazards or critical issues to demand quick action. Through Monitor Trends—Monitoring patterns, trends and public opinions by analysing sentiments/attitudes associated with various illnesses.

- Ensuring Data Privacy as Ethical Considerations: Ensuring patient data used for Sentiment Analysis is managed with utmost care, confidentiality, privacy and under applicable healthcare regulations. Data Security—Employing stringent data security protocols to safeguard patient information and avoid breaches.
- Integration with Electronic Health Records (EHR) for Patient Narrative Analysis: Exploring the integration of Sentiment Analysis into Electronic Health Records (EHR) to analyse patient's narratives and gain an in-depth insight about the patient experiences.
- Evaluation of Public Health Campaigns: Analysing people's views regarding health awareness campaigns and initiatives in public health awareness. Effectiveness Assessment—Evaluating the efficacy of public health messaging and running campaigns via Sentiment Analysis.
- Patent Education: Evaluating the opinions associated with patient's education data to ensure they are well received and understood accordingly.
- Improving Information Dissemination: Improving the dissemination of health information to patients by using insights.

Sentiment Analysis in the healthcare sector provides insightful that can be used to enhance patient experiences, improve healthcare services, and guide public health strategies. However, it is of crucial to approach the implementation with consideration of privacy, ethical principles, and the legal framework of the healthcare industry.

2.2.5.2 Advantage of Sentiment Analysis in the Healthcare Sector

The application of Sentiment Analysis in the healthcare and wellness industry has several advantages that contribute to the overall improvement of patient care, enhanced operational efficiency, and a more comprehensive understanding of entire public health dynamics.

Some significant advantages of employing Sentiment Analysis in the healthcare industry are given below:

- Identifying Areas for Quality Improvement: Through the analysis of emotions/ feelings conveyed in patient's feedback, expressed or non-expressed cues, the healthcare organizations can identify precise areas for improvement in services, facilities, medications, and patient care.
- Operational Efficiency: Operational efficiency is measured by analysing the sentiments from internal feedback/captured data to find out the morale and satisfaction levels of healthcare personnel. This information is very important for enhancing staff engagement and overall efficiency in operations.
- Feedback Analysis Based on Patient Experience Enhancement: Sentiment Analysis enables healthcare providers, integrators and hospitals, to analyse patient feedback from many sources and point of contacts like public surveys, social media, and online reviews. This observation facilitates overall understanding of patient experiences and enables addressing concerns promptly.

- Early Detection of Outbreaks Ensuring Public Health Surveillance: Monitoring sentiments on multiple social media platforms ensures early detection of pandemic, potential disease outbreaks or severe public health issues and facilitating rapid action, follow up, quick solutions and further control to block its expansion in new areas.
- Recruitment and Retention for Patient Engagement in Clinical Trials: Sentiment Analysis assists in understanding patient's emotions/feelings during pre- and post-clinical trials. It helps in improving recruitment techniques and facilities to patient's retention techniques to serve in a better way.
- Crisis Communication Planning in Public Health: At the time of severe health crises, pandemic, casualties, etc. Sentiment Analysis plays very crucial role in making best communication strategies by understanding overall public sentiments, concerns, issues, redressals and corrective measures against reactions.
- Telehealth Services Optimization: Enhancing interactive and adaptive experience of all stakeholders via Sentiment Analysis. It plays important role in capturing relevant data to understand sentiments of the patients for arranging, facilitating, providing, enhancing experience and service delivery in a best possible way while rendering telehealth services by the service providers or the concerned stakeholders.
- Brand Monitoring, Perception, and Marketing: Sentiment Analysis enables healthcare providers and other critical stakeholders to monitor the brand perception and assess the influence of promotional efforts on overall public sentiment.
- Content Evaluation for Providing Best Patient Education: Sentiment Analysis helps in assessing, modifying, and updating the contents to ensure the efficacy of patient education materials. It is also pertinent to ensure that the content is well-received, easily understood and proper feedback provided.
- Decision Support for Healthcare Providers: Decision support for healthcare providers discusses utilizing Sentiment Analysis to get actionable insights, envision further course of action, enabling data-driven decision-making and continual improvement of services.
- Sentiment Monitoring Ensuring Disease Management and Awareness: Analysing sentiments relating to specific diseases that is crucial for disease management, disease prevention and awareness campaigns to do list for due safeguard against such diseases.
- Detection of Adverse Event: Monitoring medication sentiments by analysing the sentiments associated with pharmaceuticals. It helps to provide aids well in time on account of early identification of any negative occurrences, emergencies, or potential issues with the medications.
- Data-Driven Policy Making and Its Impact: Conducting Sentiment Analysis on overall public opinions regarding healthcare policies, changes on government aids, and its impact on all stakeholders enables policymakers to understand the implications and ramifications of policies on public perception.
- Public Health Campaign Evaluation and Its Effectiveness: Sentiment Analysis analyses the efficacy of public health initiatives, ongoing campaigns, and adaptive

evaluations for providing genuine feedback for changes in ongoing campaigns as well as helping in future campaign planning.

- Patient-Centric Care Ensuring Personalized Healthcare: By understanding the individual sentiments of each patient, feedback and corrective measures can be taken to address issues raised by the patient. It facilitates healthcare providers and other stakeholders to customize services according with patient preferences and deliver a more patient-centric approach to care on priority basis within a timeframe.
- Strategic Planning and Proactive Decision: Sentiment Analysis enhances strategic planning and helps to take proactive decisions by providing useful insights into overall public views, allowing healthcare organizations and other important stakeholders to align their strategies with the demands of the public needs as detected from the logs and algorithms.
- Early Warning System: Sentiment Analysis works as an early warning system by detecting concerns, finding patterns and related issues as expressed by patients or the captured overall public sentiments/opinions for enhanced protective system.
- Ethical and Privacy Considerations: In the process of using Sentiment Analysis and studying its efficacies, the healthcare providers/stakeholder must follow required ethical norms to protect confidentiality, privacy, and secrecy of their data on priority basis.
- Monitoring Healthcare Trends: Gaining Insight into Public Health Trends-Sentiment Analysis helps in understanding public feelings regarding different healthcare trends, enabling proactive actions.

Sentiment Analysis in healthcare is a flexible technique that enables healthcare organizations to actively listen to patients, understand public attitudes, and make data-driven decisions that enhance overall healthcare delivery. When employed with ethical and responsible practices, Sentiment Analysis significantly contributes to the improving healthcare services and outcomes.

2.2.6 *Sentiment Analysis in Politics and Public Opinion [8, 9]*

Politics and public opinion are closely related. Using Sentiment Analysis in the domain of politics and public opinion provides valuable insight into the attitudes and opinions of the public towards political people, parties, policies, and societal issues.

Application: Sentiment Analysis is utilized in political campaigns to understand public opinion about candidates, parties, and policies.

Example: Analysing social media dialogues to assess popular opinion during an election campaign.

2.2.6.1 Methods to Use SA in Politics and Public Opinion [10]

- Social Media Monitoring: Apply Sentiment Analysis methods to monitor social media platforms, forums, and online discussions for mentions about political figures, parties, or specific issues.
  Example: Obtain real-time understanding of public opinions, identify emerging patterns, and evaluate overall sentiment landscape.
- New and Media Analysis: Apply Sentiment Analysis on articles, op-eds, and media coverage about political events.
  Example: Acquire knowledge about the impact of media representations on public sentiments and understand the differing contributions of different media sources in moulding viewpoints.
- Political Speech Analysis: Analyse the sentiments expressed in political speeches, interviews, and public addresses.
  Example: Evaluate the impact of political communications among the audience, identify key issues, and evaluate the effectiveness of communication strategy.
- Election Campaign Monitoring: Incorporating and integrating Sentiment Analysis into political campaigns to assess, evaluate and find patterns related to public responses in view of profile of the candidates, their speeches and critical election and campaign events.
  Example: Getting insight into the change of public sentiment throughout the election, campaign, and result by identifying possible areas of improvement, further modifications and adaptive methods to ensure better outcomes accordingly.
- Public Sentiment Tracking: Employ continuous sentiment monitoring to observe the progression of public sentiments over time.
  Example: Detecting changes in overall public opinion, uncovering enduring patterns, and evaluating the influences of events on general mood as per public opinion.
- Public Reaction to Events: Employ Sentiment Analysis to assess the prevailing public sentiment in reaction to significant occurrences, emergencies, or political advancements.
  Example: Getting valuable insights into the impact of events on public sentiment, identifying pinpoint areas of worry, and reacting immediately to emergent issues as and when they occur.
- Issues-Based Analysis: Applying Sentiment Analysis to assess, evaluate, and apply corrective steps based on the sentiments expressed in conversations about political parties, matters or policies.
  Example: Identifying public sentiments regarding significant policy matters and evaluating the impact of policy choices on overall public opinion and customize communication methods appropriately.

2.2.6.2   Advantage of SA in Politics and Public Opinion

- Real-Time Decision Making: Political parties, their campaigns, and their organizational set-up can quickly react on account of changing public sentiments/opinions. It helps in taking adapting strategies on real time basis.
- Issue Identification: Identifying important and burning issues that strongly identified/associated with people in general and enabling policymakers to effectively prioritize and efficiently address them.
- Campaign Strategy Optimization: Enhance political campaign plans by utilizing real-time Sentiment Analysis to ensure message alignment with public emotions.
- Public Engagement: Improve engagement with the public by understanding and dealing with public sentiments, cultivating a bond between politicians and the electorate.
- Policy Assessment: Evaluate public sentiments regarding specific policies, so enabling policymakers to make well-informed decision and adapt policy strategies in response to public input.
- Crisis Management: Minimize possible crises by carefully monitoring public sentiments and promptly resolving issues.
- Political Communication Effectiveness: Evaluate the efficacy of political communication techniques by understanding the reaction and interpretation of communications by the larger audience.
- Public Perception Management: Public Perception Management is the strategic shaping and management of how the public perceives a particular entity or issue. This is achieved by strategically designing communication techniques to coincide with the attitudes and values of the public.

Sentiment Analysis in politics and public opinion offers a data-centric method for understanding and responding to the constantly changing environment of public views. This subsequently enhances political strategies and facilitates well-informed decision-making.

2.2.7   Sentiment Analysis in E-Learning and Education [9, 11, 12]

The use of Sentiment Analysis within the fields of e-learning and education can provide several advantages through the providing of insight on student sentiments, levels of engagement, and overall content satisfaction with educational materials.

- Application: Sentiment Analysis is used in education sector to analyse student feedback, forum discussions, and social media chats with the aim of understanding student sentiments and improving the overall learning experience.
- Example: Analysing student reviews of online courses to improve course content and delivery.

2.2.7.1 Methods to Use Sentiment Analysis in E-Learning Education

- Learner Feed Back Analysis: Implement Sentiment Analysis into e-learning platforms student feedback forms, surveys, and comments.
  Example: Obtaining valuable insights regarding learner's evaluations of course quality, educational materials and overall learning experiences. Identifying areas that require development and revising the course material accordingly.
- Forum and Discussion Sentiment Analysis: Constructing an analysis of sentiments expressed in discussion boards and online forums included in the e-learning platform.
  Example: Gaining valuable insights into learner's engagement, identifying frequent challenges and cultivating a positive educational atmosphere through the solution of concerns and promotion of positive interactions detected from sentiments/opinions expressed.
- Personalized Learning Pathways [13]: Applying Sentiment Analysis to decipher insights along with the unique preferences and learning styles of individual learner.
  Example: Customizing paths to learning according to the individual sentiment profiles and preferences, thereby delivering an extremely personalized and effective educational experience.
- Assessments and Quizzes Feedback: Applying Sentiment Analysis on learner's comments and sentiment/opinion/feedback obtained after assessments and quizzes.
  Example: The ability to identify challenging areas in assessments, evaluating the effectiveness of questions, and determining the overall level of satisfaction with the evaluation process.
- Course Content Relevance Analysis: Evaluating sentiments about specific educational resources, lectures, training sessions, seminars, workshops, or modules.
  Example: Maintaining the relevance, updation, and value of the content. Determine which areas may require further explanation or additional information.
- Instructor Evaluation: Employing Sentiment Analysis to evaluate learner's feedback about instructors, teachers, trainers, and coaches along with methods of instruction.
  Example: Getting insight into learners' perceptions of instructors/teachers/trainers/coaches, identifying effective teaching/instructing strategies that elicit positive responses, and resolve concerns about teaching approaches.
- Emotional Engagement Monitoring: Analysing sentiment expressed or captured in open-ended responses or remarks to find out emotional engagements. Facilitating understanding of learners' emotional states during the learning process, thereby contributing to the development of a more engaging, effective, and supportive educational environment.

2.2.7.2 Advantages of Sentiment Analysis in E-Learning Education [14, 15]

- Continuous Improvement: Utilizing real-time feedback to enhance course content, instructional methods, and overall learning experiences.
- Identifying At-Risk Learners: Early identification of learners who experience negative emotions can facilitate proactive interventions aimed at providing required support to those who are at increased risk of disengaging or dropping out.
- Enhanced Learning Analytics: Incorporating and integrating sentiment data into supplement traditional educational analytics to obtain a more comprehensive understanding of the learner's experience, learning journey and learning outcomes.
- Leaner Engagement Enhancement: Cultivating a constructive environment for learning through the active handling of learner's questions, getting learner's interest into the subject/topic/type of study materials, promoting constructive dialogues and modifying materials to optimize overall learner's engagement and development.
- Enhanced Learner's Satisfaction: Encouraging a positive learning environment, resolving all upcoming issues during learning process, and adapting contents by individual preferences and selection to improve overall learner's satisfaction.
- Evaluating Instructor's Effectiveness: Conducting an evaluation of the instructors' effectiveness and hence usefulness to identify core areas that require improvement to enhance the overall quality of instructions. Adaptive learning can be achieved by customizing assignments, the types of study material, audio/video/applications/case studies/logical analogy, speed of deliverance of the study materials and quality of the contents according to the sentiment profiles of individual's learner. It will promote a more individualized learning environment.

By integrating and collaborating Sentiment Analysis into e-learning and education, organizations may harness the insights of learners to develop learning experiences that are more engaging, efficient, effective, and individualized. The implementation of a data-driven approach enables continuous improvement and provides long-term benefits in terms of learner's achievement and satisfaction.

2.2.8 Sentiment Analysis in Human Resources [3]

In the area of Human Resources (HR), integrating and collaborating the use of Sentiment Analysis provide significant valuable insights of the employee/employer sentiments, their emotions on agreements/differences over many issues, job satisfaction and employer's assessment and overall engagement for betterment of the employee–employer relationship.

The advantages of integrating Sentiment Analysis in human resources are as follows:

- Application: Human resources use Sentiment Analysis to assess employee's requirements, satisfaction, suitability of the role, evaluation based on feedback/ emotions, and overall HR management. It helps in identification of the potential of the employee as well as how organization to make planning for maximizing productivity and addressing issues.
- Example: Recognizing areas for improvement, critical issues and challenges and understanding overall job satisfaction by analysing employee survey including responses link to sentiments.

2.2.8.1 Methods to Use Human Resources Sentiment Analysis

- Employee Surveys and Feedback: Apply Sentiment Analysis on comments, feedback forms, behavioural changes, responses on social media platforms and employee survey responses.
  Example: Getting valuable insight into the overall perception and personality of the staff members regarding policies, workmanship, team-spirit, discipline, leadership, proactive, attitude to serve customers, friendliness among colleagues, and compatibility with the culture of the organization. It helps to address specific areas for further improvement and fine-tuning.
- Performance Evaluations: Analysing the sentiments conveyed in performance evaluations through various assignments, applications, case studies, tests, quizzes, debates, lectures and the feedback provided by senior managers, colleagues and subordinates.
  Example: Acquiring valuable insights about employees' perceptions of their performance evaluations and alignment with corporate objective and its culture. Identifying patterns, trends, or domains that could benefit for improvements based on feedback/opinion.
- Training and Development Feedback: Conducting a Sentiment Analysis of feedback/opinion received before and after the professional development programs or training sessions to understand the gap and taking corrective measures.
  Example: Get an understanding of how employees evaluate the efficacy and back home utility of the training endeavours. Tailor-made training programmes to the designed to fulfil the requirements and preferences of employees by utilizing those insights into workable actions.
- Employee Communication: Conducting Sentiment Analysis on internal communication channels, including but not limited to emails, intranet messages, social media posts, publishing articles, preparing office notes/drafting and team collaborative tools.
  Example: The ability to monitor employee's reaction to all such internal communications, updates, engagements and announcements. Adapting strategies for faster and better communication in relation to sentiment feedback.

- Surveys of Employee Engagement Surveys: Incorporating and integrating Sentiment Analysis into personal surveys by measuring employee engagement and satisfaction. Achieving an increased awareness of the sentimental attachments exhibited by personnel or captured by machines logs/algorithms.
  Example: Determining the factors influencing job satisfaction and identifying potential areas for enhancement to foster overall engagement, satisfaction, and improvement.
- Diversity Inclusion: Implement Sentiment Analysis on the opinions of employees related to activities designed to foster diversity and inclusion.
  Example: Understand how diversity and inclusion initiatives influence the experiences of employees. Determine sentiments about the environment of diversity and inclusiveness within the organization.
- Exit Interviews: Implementing Sentiment Analysis on the responses of the employee obtained from the exit interviews, feedback on socials media sites, company's websites etc.
  Example: Recognizing common sentiments expressed by majority of the employees who are leaving their positions and the company. Applying this knowledge to resolve problems and enhancing employee retention strategies.

2.2.8.2 Advantages of Human Resources Sentiment Analysis

- Employee Satisfaction: Achieve a greater understanding of the levels of employee satisfaction. Define the domains in which personnel are satisfied and those that require improvement.
- Employee Retention: Identify the sentiments articulated by departing personnel, thereby facilitating the resolution of problems and enhancing retention strategies.
- Workplace Culture: Gain insight into the dominant perspectives concerning employee morale, organizational values, and workplace culture.
- Leadership Evaluation: Assess the sentiments expressed towards regarding management and leadership. Identify domains that require leadership growth and enhancement.
- Proactive Issue Resolution: Utilize Sentiment Analysis of diverse HR-related Enhanced Communication. Communications to identify potential HR issues in their early stages to address concerns proactively.
- Improved Communication: Adapt internal communications to align with employee sentiments, thereby ensuring that messages are well received and appreciated by the workforce.
- Implementing Data-Driven Decision Making: Instead of exclusively relying on subjective observations, base HR decisions on insights derived from data. This results in strategic and well-informed decision-making.
- Enhancing Employee Engagement: Employ Sentiment Analysis to discern the determinants that impact employee engagement. Incorporate focused strategies to augment levels of overall engagement.

By incorporating and integrating Sentiment Analysis into human resources procedures, pre and post employee assignment feedback from both employee and employer. Businesses may get benefits of Sentiment Analysis that can be cultivated in existing work environment making favourable and efficient, promptly attending to concerns, and ultimately improving the experience of the employees. This methodology is consistent with the tenets of ongoing enhancement and human resources principles and practices that prioritize employee preferences and comfort to employer to take suitable corrective measures, if required.

2.2.9 Sentiment Analysis in Tourism and Hospitality [16]

In the areas of tourism and hospitality, the application of Sentiment Analysis can serve as an effective tool for businesses to understand customer needs, pre-requisites, sentiments, improve service offerings, and manage their online reputation and facilitate prospects.

An overview of possible benefits of applying Sentiment Analysis in the tourism and hospitality sector are as follows:

- Application: The tourism industry utilizes Sentiment Analysis for analysing reviews and feedback, thus helping businesses understand customer satisfaction and make improvements.
- Example: Aspects that contributing to positive or negative experiences for guests can be identified by analysing online hotel reviews.

2.2.9.1 Methods to Use SA in the Hospitality and Tourism

- Social Media Monitoring: Monitor social media platforms and all reviews/posts for references to the hotel, restaurant, or tourism services which constitutes the process.
  Example: Real-time understanding of customer sentiments enables timely and effective responses to both positive and negative sentiments.
- Review Analysis: Analysing Sentiment Analysis on feedback from customers present on platforms like Google Reviews, TripAdvisor, or Yelp, etc.
  Example: Acquiring knowledge regarding the sentiments that are linked to various aspects of the services, including dining, accommodation, and customer service by identifying trends and areas for improvement.
- Visitor Reviews: Using Sentiment Analysis on customer satisfaction survey responses.
  Example: The ability to quantify levels of customer satisfaction and identify specific areas that contribute to positive or negative sentiments.
- Competitor Analysis: Methods: Conduct a Sentiment Analysis of competitors operating within the hospitality and tourism sector.

Example: Gain insight into the comparative performance of your business in relation to competitors, identifying both term of positive aspects to emphasize and areas that may need improvements.

- Event Analysis: Apply Sentiment Analysis during or after events organized by your business.
  Example: Understanding of the sentiments of attendees and identify positive aspects as well as areas that require improvement for subsequent events.
- Feedback from Online Platforms: Monitor the sentiments and feedback expressed on online travel forums, blogs, and community platforms.
  Example: Capture sentiments from a wide range of sources in order to obtain a comprehensive understanding of the experiences and opinions of customers.

2.2.9.2 Advantages of Sentiment Analysis in Hospitality and Tourism

- Customer Experience Improvement: Utilize Sentiment Analysis insights to enhance different aspects of the customer experience, which leads to increased levels of satisfaction and loyalty.
- Reputation Management: Implement proactive management of online reputation using prompt responses to negative sentiments and amplification of positive experiences.
- Service Improvement: Identify specific areas that could benefit from improvements, including but not limited to staff behaviour, cleanliness, or amenities, therefore leading in a more improved offering as a whole.
- Marketing Strategy Optimization: Adjust marketing strategies under customer sentiments to emphasize aspects that create a positive favourable response.
- Real-Time Decision Making: Response to customer sentiments in real-time, which include during promotional activities, events, and crisis situations.
- Competitive Advantage: Acquiring competitive edge by understanding the ways in which the organization separates itself or falls short about consumer sentiments or employee's engagement when compared to competitors and their strengths and weaknesses. Customer engagement means actively participating in conversations with consumers on social media platforms, responding to negative sentiments with reassurance and expressing gratitude for positive feedback or further improvement.
- Trend Prediction: Identifying emerging trends in consumer preferences and latest innovations related to the products and services and expectations to maintain a competitive edge for your organization and assurance to generate revenue in future as well in view of sustainable growth.

The adoption of Sentiment Analysis in hospitality and tourism sectors will promote a customer-centric, data-driven and solution providing approach. It enables organizations to garner a lot of valuable information, how to incorporate adaptive changes and customize services in response to prompt consumer feedback, ultimately leading to improved customer satisfaction and business success.

2.2.10 Sentiment Analysis in Production Developments [17]

The application of Sentiment Analysis to the field of product development can provide significant insights into the customers opinions, preferences, and expectations.

The benefits of incorporating Sentiment Analysis into product development are as follows:

- Application: Businesses use Sentiment Analysis to gain insight into consumer preferences and expectations, enabling them to create new products.
- Example: Analysing consumer feedback on social media platforms in order to identify the specific features that they prefer to future software release.

2.2.10.1 Methods to Utilize SA in the Product Development

- Customer Feedback Analysis: Sentiment Analysis is applied to customer feedback, evaluations, understanding the value of the feedback vis-a-vis profile of the customer and remarks regarding established products and services or query about future demands.
  Example: Getting insight into potential product enhancements, identifying strengths and weaknesses, and comprehending how consumers currently perceive existing offerings and what more expect.
- Feature Sentiment Analysis: Analysing sentiments related to specific features or attributes of a products/services.
  Example: Acquiring knowledge regarding the customer-centric features that might need further improvement or modification in future iterations.
- Competitor Product Analysis: Expanding Sentiment Analysis to reviews, validations and feedback evaluations about other competitor's products.
  Example: Getting insight into customer's needs and dislikes regarding competitors' products, critical benchmarking about value enhancement of product/service.
- Pre-launch Analysis: Sentiment Analysis for a new product or feature, utilizing market research, surveys and social media discussions as sources of information.
  Example: Assessing levels of anticipation, identify possible concerns, reality check from markets, and adjust marketing strategies or the product itself before and after its launch as per market dynamics.
- Post-deployment Evaluation: Apply Sentiment Analysis on post-product-launch consumer reviews, social media comments, feedback and further scope of development.
  Example: Evaluating the product's overall reception, identifying areas that require improvement, and formulating decisions regarding future product iterations based on empirical/static as well as moving data.
- The Identification of Product Gaps: Analysing sentiments related to consumer preferences and expectations that current products might fail to fulfil.

Example: Identifying market gaps or unfulfilled consumer demands, thus producing ideas for the creation of new products.
- Brand Loyalty Analysis: Analysing sentiments about attachment, perception and loyalty about the brand.
  Example: Gaining insight into the emotional connection's customers have with the brand and products; this knowledge can guide choices regarding maintaining or adapting brand equity and its identity for customers.
- Social Media Listening: Monitoring sentiments on social media platforms that relate to discussions about the products/services or industry.
  Example: Tableting data about well-informed value statements regarding emerging trends, customer sentiments, customer preferences, and potential issues that may impact product/service development decisions.

2.2.10.2 Advantages of SA in Product Development

- Data-Driven Decision-Making: Implementing a data-driven approach to add value in decision-making by utilizing real-time information to identify customer sentiments, preferences, queries, concerns, criticisms and commendations.
- Enhanced Product Quality: Improving product quality by addressing specific issues, product acceptability/rejection, high demand of the product and weaknesses, if any identified or expressed through Sentiment Analysis of consumer feedback.
- Innovation Inspiration: Identifying area for innovative/market driven product development and creating guidelines after proper analysis about the unfulfilled customer demands or desires, if any.
- Improved Marketing Strategies: Adapting market strategies as per customer sentiments and ensuring that promotional efforts aligned with customer expectations and choices.
- Reduced Time-to-Market: It is possible to reduce the time required to introduce successful products/services to the market by identifying and resolving issues during early phases of product development.
- Competitive Advantage: Satisfying customer expectations better than competitors by understanding their sentiments and delivering products tailored to their needs to minimizing gap.
- Brand Reputation Management: Effective brand reputation and re-assurance management involves the proactive solution of consumer issues and concerns to safeguard and enhance the overall reputation of the brand in long-term perspective.

By integrating Sentiment Analysis into the product development process, organizations gain the ability to produce products and services that outperform customer expectations, market dynamics and beating competitors with a handsome margin. It thereby helps in developing brand loyalty and securing customer satisfaction for sustainability of the company.

The applications mentioned above demonstrate the adaptability of Sentiment Analysis in gathering valuable insightful information from unstructured textual/audio/video data in a variety of domains. As more organizations recognize its true potential for facilitating informed decision-making, aligned strategy development and fulfilling market demands continuously on basis of adaptation and continues upgradations.

2.3   Conclusion

In conclusion, the application of Sentiment Analysis across the diverse domains is very crucial for the transformative impact on the processes and particularly understanding human sentiments said/expressed in textual/audio/video, social media platforms and hidden patterns deciphered by data analysis. From customer feedback analysis in business to monitoring public opinion on social media, Sentiment Analysis provides invaluable insights for informed decision-making and further improvement. The education sector benefits from understanding learner's sentiments, while healthcare applications leverage Sentiment Analysis of the patient to enhance patient experiences and satisfaction. The continual advancements in natural language processing, Machine Learning, artificial intelligence and Deep Learning technologies further expand the horizons of Sentiment Analysis and there is further scope of fine-tuning and development. As we navigate through these diverse domains, it is evident that Sentiment Analysis has become an indispensable tool for businesses, institutions, and researchers alike. Its ability to decode the emotional aspects of language not only refines strategic approaches but also build a more matured understanding of human communication in the digital age. Looking forward, the evolving landscape of Sentiment Analysis promises even greater precision, adaptability, productive, acceptability, and ethical considerations in its applications across an ever-expanding array of fields of science and humanity.

References

1. Singh SK, Paul S, Kumar D (2014) Sentiment analysis approaches on different data set domain: survey. Int J Database Theory Appl 7(5):39–50. https://doi.org/10.14257/ijdta.2014.7.5.04
2. Dolianiti F, Iakovakis D, Dias S et al (2018) Sentiment analysis techniques and applications in education: a survey. In: International conference on technology and innovation in learning, teaching and education. Springer. Accessed 28 Jan 2024. [Online]. Available: https://doi.org/10.1007/978-3-030-20954-4_31
3. Wankhade M, Rao ACS, Kulkarni C (2022) A survey on sentiment analysis methods, applications, and challenges. Artif Intell Rev 55(7):5731–5780. Accessed 28 Jan 2024. [Online]. Available: https://doi.org/10.1007/s10462-022-10144-1
4. Palli AS et al (2014) A study of sentiment and trend analysis techniques for social media content. Int J Mod Educ Comput Sci 12:47–54. https://doi.org/10.5815/ijmecs.2014.12.07

5. Jain R, Kumar A, Nayyar A, Dewan K, Garg R (2023) Explaining sentiment analysis results on social media texts through visualization. Multimed Tools Appl. Accessed 27 Jan 2024. [Online]. Available: https://doi.org/10.1007/s11042-023-14432-y

6. Lou C, Atoui MA, Li X (2023) Recent deep learning models for diagnosis and health monitoring: a review of research works and future challenges. https://doi.org/10.1177/014233122 31157118

7. Hiremani V, Devadas RM, Patil H, Patil S, Sweta S, Patil V (2024) Hypokinetic rigid syndrome prognosis using random forest classifiers and support vector machines. Int J Intell Syst Appl Eng 2024(14s):632–636. Accessed 10 Feb 2024. [Online]. Available: https://ijisae.org/index.php/IJISAE/article/view/4710

8. Pang B, Lee L (2008) Opinion mining and sentiment analysis. Found Trends Inf Retr 2(1–2):1–135. https://doi.org/10.1561/1500000011

9. Shaik T, Tao X, Dann C, Xie H, Li Y (2023) Sentiment analysis and opinion mining on educational data: a survey. Nat Lang Process J. Accessed 27 Jan 2024. [Online]. Available: https://www.sciencedirect.com/science/article/pii/S2949719122000036

10. Sonia, Sharma K, Bajaj M (2022) A review on opinion leader detection and its applications. In: 7th international conference on communication and electronics systems, ICCES 2022—proceedings. Institute of Electrical and Electronics Engineers Inc., pp 1645–1651. https://doi.org/10.1109/ICCES54183.2022.9835870

11. Sweta S (2021) Educational data mining in e-learning system. In: Modern approach to educational data mining and its applications. Springer, pp 1–12. https://doi.org/10.1007/978-981-33-4681-9_1

12. Sindhu I, Muhammad Daudpota S, Badar K, Bakhtyar M, Baber J, Nurunnabi M (2019) Aspect-based opinion mining on student's feedback for faculty teaching performance evaluation. IEEE Access 7:108729–108741. https://doi.org/10.1109/ACCESS.2019.2928872

13. Sweta S, Lal K (2017) Personalized adaptive learner model in E-learning system using FCM and fuzzy inference system. Int J Fuzzy Syst 19(4):1249–1260. https://doi.org/10.1007/S40815-017-0309-Y

14. Sweta S (2021) Educational data mining techniques with modern approach. In: Springer briefs in applied sciences and technology, pp 25–38. https://doi.org/10.1007/978-981-33-4681-9_3

15. Sweta S (2021) Modern approach to educational data mining and its applications. Accessed 29 Jan 2024. [Online]. Available: https://doi.org/10.1007/978-981-33-4681-9.pdf

16. Medhat W, Hassan A, Korashy H (2014) Sentiment analysis algorithms and applications: a survey. Ain Shams Eng J. Accessed 28 Jan 2024. [Online]. Available: https://www.sciencedirect.com/science/article/pii/S2090447914000550

17. Pathak AR, Agarwal B, Pandey M, Rautaray S (2020) Application of deep learning approaches for sentiment analysis, pp 1–31. https://doi.org/10.1007/978-981-15-1216-2_1

Chapter 3
The Transformative Role of Sentiment Analysis in Education

3.1 Introduction

The study of Sentiment Analysis has opened new study in the domain of Educational Data Mining. It is an important tool to analyse the captured emotions, incidents and valuable inputs while the learner was interacting and integrating with the educational system. It helps in adapting the learners in the learning environment and ensures overall improvement to learner as well as the education system. It also enhances value of learning experiences efficiently and effectively by adding some more dimensions. Sentiment Analysis may be linked with opinion mining or behaviour mining in general terms to understand the requirements of the learners in a most suitable way. It is not only limited to the academic excellence, but also goes beyond to enrich all stakeholders by offering an in-depth understanding of learner's sentiment that can be most important in making educational strategy to improvise the entire education system. This chapter discusses the significant impacts of Sentiment Analysis in education through EDM, i.e. analysing its uses, advantages, concerns, and way forwards.

Sentiment Analysis is the synthesis of the application of NLP, machine learning. Deep learning and computational linguistics to evaluate, assess, and find out the sentiments expressed in textual/audio/video/social media/behavioural data. Within the context of the educational domain, such information may be obtained from several sources like learner's feedback, online conversations through blogs, chats and forums, written text paragraphs, verbal communications as stored in the system, behavioural cues and interaction on social media platforms.

S. Sweta, *Sentiment Analysis and its Application in Educational Data Mining*,
SpringerBriefs in Computational Intelligence,
https://doi.org/10.1007/978-981-97-2474-1_3

3.2   Applications of Sentiment Analysis in the Field of Education [1–3]

Recently, the integration and collaboration of Sentiment Analysis (SA) and Educational Data Mining (EDM) has emerged as the most prominent, interesting and active area of research [1]. Currently, there are not many comprehensive literature reviews available related to this hybrid domain. As a result, this chapter focuses primarily on the powerful scientific literature reviews related to Sentiment Analysis (SA) in EDM and defines various possibilities for future research applying SA, keeping in consideration of the limited number of reviews from the engineering and technology fields. The primary stakeholders identified in these studies are educators, instructors, academics, and learners or learners. Additional studies have exposed the granularity of sentiment categories in educational data, with the discussion of hybrid approaches for SA being the significant and most closely related to it. However, in-depth investigation, identification, and discussion were devoted to the four major SA research topics including designing sentiment methods or systems, investigating learners' satisfaction/skill/attitude about the concerned topics, evaluating teachers' teaching performance and examining the relationship between various sentiments, behaviour, performance and achievement [4, 5].

- Learner's Feedback Analysis:
  Sentiment Analysis is a valuable tool to analyse learner's feedback, interactions and evaluations across the learning curve. It provides useful perspectives to educators, administrators and all other stakeholders in view of recognizing all pros and cons, weaknesses and strengths and areas for improvement in teaching pedagogy, course content, way of delivery of the content, audio/visual/kinaesthetic approaches and overall educational experiences, innovations and creativities.
- Adaptive Learning Systems:
  Sentiment Analysis can be integrated for developing adaptive learning systems to individualize instructional material along with recommender system according to the requirement and sentiments shown by the learners. This personalized system will ensure that learners may receive the contents as per their preferred way of learning skills and handle the upcoming assignments, tests, and challenges confidently to improve on all counts.
- Early Intervention and Learner Support:
  The Sentiment Analysis that revealed and captured through the learner's communications across the computational techniques, educators and policy makers may identify learners' emotional ingredients and rightly assess the academic difficulties during the learning process. Therefore, the early detection and intervention strategies may be employed to provide essential assistance in learning, developing positive vibes among learners and creating overall conducive learning environment.
- Enhanced Engagement:

Sentiment Analysis is essential to analyse the level of learner's involvement in online and interactive platforms. The learning may be imparted in offline courses also. The educators and instructors may modify their teaching or training pedagogical approaches to enhance learner's engagement and motivation by understanding the emotions of the learner during their interactive educational processes and assess how these factors are influencing in the outcome.

3.3 Advantages of Sentiment Analysis in Education [6, 7]

- Data-Driven Decision-Making System:
  Sentiment Analysis enables administrators and teachers with useful information that may be used to make tactical, strategic and well-informed decisions. Institutions can make targeted modifications to enhance the overall learning experience by understanding the sentiments and context of the learner experience.
- Enhancing Learner Satisfaction:
  Identifying and addressing concerns raised by learners leads to a sense of satisfaction. Institutions that actively engage with learner's feedback showcase their dedication to the enhancement and achievement of learner's success.
- Personalized Learning Path:
  Sentiment Analysis enables the implementation of personalized learning paths. By customizing content and preferences according to sentiments, educational institutions can provide an adaptive and learner-centric learning experience.
- Improved Teacher–Learner Relationships:
  To build relationships with learners, educators can build stronger relationships with learners by staying responsive to the sentiments conveyed by them. By understanding learner's context teachers can adopt empathetic and responsive teaching pedagogical methods.

3.4 Methods and Examples of Sentiment Analysis in Educational Data Mining [2, 8]

Sentiment Analysis in the field of data mining focuses on using techniques to analyze the sentiments conveyed in educational content, learner's feedback, and interactions with social media within educational settings.

Below are a few approaches and examples where Sentiment Analysis is applied in the context of data mining.

3.4.1 Lexicon-Based Methods

- Description: Lexicon-based methods use to find predefined dictionaries or lexicons containing words associated with sentiment scores. Based on existence of words and polarity of words, the sentiments of a piece of text are determined.
- Example: Incorporating positive and negative scores to words in learner's feedback and averaging them to determine the overall sentiment.

3.4.2 Machine Learning Methods [9]

- Description: Labelled datasets can be used to train by using Supervised Machine Learning Algorithms such as Support Vector Machines (SVM), Naive Bayes, and more advanced models such as Neural Networks, for better sentiment predictions.
- Example: Making a training model using reviews of the various learners identified as positive, negative, or neutral based on validated value propositions and mathematical algorithms and comparing the sentiment data captured from new learner with the existing trained model, new adaptive predictions can be made for the learner as well as other stakeholders.

3.4.3 Analysis of Aspect-Based Sentiment [10, 11]

- Description: This approach focuses on the Sentiment Analysis of specific aspects or subjects specified in each text. This feature proves to be especially useful in academic environments where opinions may be made regarding various aspects such as teaching and course quality or facilities.
- Example: Analysing learner's reviews to identify sentiments related to the quality of education, course materials, and extracurricular activities.

3.4.4 Deep Learning Methods [12, 13]

- Abstract: Recurrent Neural Networks (RNNs) and transformer-based models, which are based on Deep Learning architectures, have proven effectiveness in extracting sentiment expressions and contextual information.
- Example: We can perform Sentiment Analysis by using a pre-trained transformer model (e.g. BERT) and fine-tuned on educational data to perform Sentiment Analysis.

3.4.5   Rule-Based Methods

- Description: Sentiments can be indicated by identifying patterns based on pre-defined rules like linguistic features, grammatical structures, performance evaluation processes, emotional, behavioural, or specific keywords.
- Example: Identifying positive sentiments by words like "excellent," "great," or negative sentiments by "poor" or "disappointing."

3.4.6   Ensemble Methods

- Description: Ensemble methods help in enhancing overall accuracy and robustness by combining all relevant predictions from multiple Sentiment Analysis models.
- Example: Predictions from a lexicon-based model, a Machine Learning model and a Deep Learning model may be combined to get a more comprehensive Sentiment Analysis.

3.5   Some Important and Leading Examples of Sentiment Analysis

3.5.1   Learner's Feedback Analysis [14]

- Method: By using a sentiment classifier trained on a dataset of labelled learner's feedback—a Machine Learning approach.
- Example: Analysing learner's reviews of a course to identify factors influenced by sentiments about the teaching style, course materials, quality of materials, presentation of the materials, performance in tests and overall satisfaction.

3.5.2   Discussion Forum/Chat Room Sentiment Analysis

- Method: Rule-based approach to identify sentiment details in learner's posts on an educational discussion forum/chat room.
- Example: Analysing sentiments expressed by learner in forum discussions/group chat/group activities to understand their engagement, involvement, and concerns.

3.5.3 Course Content Evaluation

- Method: Aspect-based Sentiment Analysis to assess, evaluate and identify sentiments related to specific topics within educational content.
- Example: Assessing sentiments shown or expressed about the difficulty level, relevance, quality and quantity of course materials, and clarity of course materials.

3.5.4 Social Media Sentiment Monitoring [15]

- Method: Lexicon-based approach to monitor from the written or spoken expressed sentiments on social media platforms related to educational events, course materials, teacher's feedback, or institution's facilities.
- Example: Analysing tweets, re-tweets, sharing, posting, liking, and commenting about an educational conference, academic activities, teacher's efficiency, facilities at institutions, etc. for monitoring the overall sentiment of learners.

3.5.5 Feedback Forms Analysis

- Method: By combining Machine Learning and lexicon-based models to analyze sentiments in open-ended feedback forms to identify learner's characteristics.
- Example: Processing feedback forms to understand sentiments shown or expressed by the learners in a training program.
- Customization: Educational contents may have domain-specific language, cultural inclination, government's influences, quick updating, latest changes. Therefore, the customization of sentiment models is often necessary to cope up with new information.
- Data Quality: The effectiveness of Sentiment Analysis mainly depends upon the quality and relevance of the labeled training data.
- Ethical Considerations: Ensuring confidentiality, privacy, and ethical use of Sentiment Analysis results, particularly when dealing with learner's data or sharing personal data and its outcome to others.

In Educational Data Mining, Sentiment Analysis provides valuable insights into learner's experiences, engagements, involvements, feedback, concerns and areas of further improvement, contributing to the overall enhancement of educational processes, quality of the course contents and outcomes as overall improvement in efficiency and effectiveness of the educational institutions.

3.5.6 Adaptive Learning Systems [19]

- Purpose: Personalizing the learning experience based on learner's sentiments and preferences.
- Method: Integrating and collaborating Sentiment Analysis to understand how learners respond emotionally to different learning materials, allowing adaptive systems to tailor made contents can be provided to them accordingly.

3.5.7 Early Identification of Issues

- Purpose: Identifying early signs of learner's dissatisfaction or potential problems.
- Method: Implementing Sentiment Analysis on various data sources to detect positive, neutral or negative sentiments or concerns, enabling prompt intervention to enhance value.

3.5.8 Online Course Improvement

- Purpose: Enhancing online courses based on learner's knowledge, skills, attitude, behaviour, sentiments, and feedback.
- Method: Analysing sentiments in course reviews in addition to the explicit performance, discussion in forums, chat rooms, or social media's responses, reactions and emoji to identify areas that need further improvement and inform course designer to change adaptively.

3.5.9 Educational Chatbots and Virtual Assistants

- Purpose: Enhancing the interaction experience between learners and educational chatbots.
- Method: Integrating and collaborating Sentiment Analysis to understand the emotional tone of learner's queries, concerns, and responses, allowing chatbots to respond more appropriately.

3.5.10 Teacher and Instructor Feedback

- Purpose: Analysing sentiments in assessment, evaluations, and feedback provided by learners about instructors.
- Method: Applying Sentiment Analysis to instructor or teacher's reviews and feedback forms to understand how learner's perceive teaching styles and effectiveness.

3.5.11   Learning Analytics

- Purpose: Integrating Sentiment Analysis into broader learning analytics initiatives.
- Method: Utilizing sentiment and its valuable insights as one of the factors in learning analytics models to understand the holistic learner's experience and engagement.

3.5.12   Sentiment-Driven Interventions

- Purpose: Implementing interventions based on Sentiment Analysis and its outcomes to address learner's concerns and relevant issues.
- Method: Using sentiment data to identify and specify which specific areas of concern requires improvement and providing tailor-made solutions on positive interventions to enhance the overall learning experiences.

3.5.13   Emotion-Aware Learning Environments

- Purpose: Creating learning environments that respond to learners' emotional states.
- Method: Integrating and collaborating Sentiment Analysis to detect emotions and adapt the learning environment or provide additional support when they exhibit signs of frustration, failure or disengagement.

3.6   Integration of Sentiment Analysis with Educational Data Mining

Implementing Sentiment Analysis in Educational Data Mining involves selecting appropriate methods, database, variables, and algorithms based on the characteristics of the data, processing mechanisms of data, and the objective of the analysis. There are various methods and algorithms commonly used in the Sentiment Analysis for Educational Data Mining.

3.6.1   Lexicon-Based Approaches

- Description: Lexicon-based methods based on predefined sentiment lexicons or dictionaries containing words associated with the sentiment scores. The sentiment of a piece of text is determined based on the presence, priority, and polarity of these words.

- Application in Education: Lexicon-based methods are suitable for analysing sentiment in educational content, study materials, quality or quantity of the study materials, learner's feedback, or discussions.

3.6.2 Machine Learning Approaches [9, 16]

- Description: Supervised Machine Learning Models are trained and predefined on labelled datasets to predict sentiment in unseen data besides the existing one. Common algorithms include:

  - Naive Bayes: Simple probabilistic classifier based on Bayes' theorem.
  - Support Vector Machines (SVM): Classifies data points by finding the hyperplane and how the best separated from the different classes.
  - Random Forest: Combining learning method that constructs multiple decision trees.

- Application in Education: Machine Learning approaches are effective tools when dealing with large volumes of textual/audio/video datasets, such as learner's opinion/feedback, where the model learns patterns from labelled examples.

3.6.3 Deep Learning Approaches [12, 17, 18]

- Description: Deep Learning models, especially the Neural Networks with multiple layers, are used for finding patterns through Sentiment Analysis. Features of common architectures are given below:

  - Recurrent Neural Networks (RNNs): Processing sequences of data which suitable for tasks involving sequential information.
  - Long Short-Term Memory Networks (LSTMs): A type of RNN designed to capture data variables related to long-term dependencies.
  - Transformer-Based Models (e.g., BERT): Attention-based models that consider the entire context of a sequence in one go.

- Application in Education: Deep Learning approaches excel at capturing contextual variations in language, making them more suitable for Sentiment Analysis tasks in educational domain.

3.6.4 Aspect-Based Sentiment Analysis Approach [10]

- Description: This approach focuses on analysing sentiments related to specific aspects or topics within a piece of text or datasets captured from the educational processes. It allows for a more fine-grained and fine-tuned analysis of sentiments.
- Application in Education: Useful for breaking down sentiments into further finer categories like analysing sentiments about teaching quality, quality/content of the course materials, relevant course materials or administrative processes separately.

3.6.5 Rule-Based Approaches

- Description: Rule-based Sentiment Analysis based on predefined rules or patterns to determine sentiment from the text/audio/video, etc. Rules can be based on linguistic features, grammatical structures, qualification criteria in tests or specific keywords.
- Application in Education: Rule-based approaches are useful for specific contexts where sentiment can be identified through clear linguistic cues exhibited or clearly expressed. For example, identifying sentiments in survey responses directly or taken forum posts after analysing.

3.6.6 Hybrid Approaches

- Description: Hybrid approaches combine multiple methods like integrating lexicon-based analysis with machine learning models or combining rule-based and deep learning methods.
- Application in Education: Hybrid approaches aim to leverage the strengths of different techniques to improve overall accuracy and adaptability across diverse educational data.

3.6.7 Ensemble Methods Approach

- Description: Ensemble methods combine predictions from multiple models to enhance overall accuracy, effectiveness and robustness.
- Application in Education: Combining predictions from different Sentiment Analysis models like lexicon-based, Machine Learning, neural networks, and Deep Learning models. It can provide a more comprehensive Sentiment Analysis.

3.6.8 Transfer Learning

- Description: Transfer learning involves pre-training a model on a large dataset and fine-tuning it for a specific task. This is particularly relevant in the context of Deep Learning and Machine Learning models.
- Application in Education: Pre-trained models can be fine-tuned on educational data to improve performance and adapt to the specific languages used in the educational context for overall improving effectiveness.

3.6.9 Sentiment Analysis on Social Media

- Description: Analysing sentiments expressed on social media platforms related to educational institutions, courses, subjects or events.
- Application in Education: Social Media Sentiment Analysis is valuable for understanding public perception, identifying trends, patterns and monitoring sentiments about educational entities and all stakeholders.

3.6.10 Temporal Analysis

- Description: Analysing changes in sentiments over time, which is particularly relevant in educational contexts and the analysis to be used by all policy-makers and educators as sentiments are critical factors in finalizing academic calendars, events, courses, or policy changes.
- Application in Education: Temporal analysis helps in identifying trends, patterns and understanding how sentiments fluctuate in response to various factors and influence in totality.

These methods, logs, and algorithms can be adapted and combined based on the specific requirements, new assessments and variations of Sentiment Analysis in the Educational Data Mining context specifically or data driven analysis. The choice of the approach depends on factors like the nature of the data, the scale of the analysis, applicability of the data analysis and the specific goals of Sentiment Analysis within the educational context.

As a conclusion, Sentiment Analysis in Educational Data Mining plays a pivotal role in understanding, identifying and enhancing the learner's experience, improving instructional design, drafting adaptive policies or courses and fostering a positive and effective learning environment.

3.7 Challenges and Considerations in Sentiment Analysis

As Sentiment Analysis holds immense promises; so, there are some challenges to surmount to get better results. Ethical considerations, privacy concerns, confidentiality, permission from the stakeholders to share the data and results with others and the need for accurate sentiment interpretation are among the important aspects educators, policymakers, and institutions must carefully navigate. Some considerations and challenges are given as under:

- Privacy: Ensuring that Sentiment Analysis is conducted in compliance with privacy and confidentiality regulations, especially when dealing with personal data.
- Bias: Being mindful of potential biases in Sentiment Analysis models, identifying them and addressing them to ensure fair to all and accurate results for better effectiveness of analysis.
- Integration: Seamlessly integrating and collaborating Sentiment Analysis into Educational Data Mining processes, datasets, variations, outcomes, and compatible platforms for efficient and effective uses.

3.8 Future Implications

As technology continues to evolve, the future of Sentiment Analysis in education holds exciting possibilities. The integration of artificial intelligence, advanced Machine Learning models, Deep Learning, neural networks, artificial intelligence, and cross-domain Sentiment Analysis could further fine-tune the valuable insights derived from learner's sentiments.

3.9 Conclusion

The transformative role of Sentiment Analysis in education is remarkable and its effectiveness to enhance overall education system can't be ruled out. Our educational systems navigate the digital age, AI based high end technological revolution and changing dynamics of learners and learning process. Therefore, understanding and harnessing the sentiments expressed by learners, educators, policy makers and other stakeholders become most important aspects for overall improvement in education system. Sentiment Analysis emerges as a powerful tool, offering different lens to capture the entire educational journey into the emotional landscape and the given educational environment and the ecosystem. By decoding sentiments within learner's feedback, discussions, opinion, feedback from peers, deciphered information from social media chats, online forums, and online interactions, educational institutions gain valuable insights that can enhance strategic decision-making processes, enhance teaching methodologies/styles/pedagogies, proactive arrangements by educators, facilitating adaptivity and improvise overall learner's satisfaction along with win–win situation for all other stakeholders. The ability of Sentiment Analysis to unveil trends, understand patterns, decipher important information, critical analysis to use them prudently, incorporate and collaborate adaptive changes, detect potential issues, future challenges, way forward to address issues and challenges and spotlight or sweet areas of excellence making it an indispensable tool for shaping a responsive and learner-centric educational landscape fostering futuristic goals. As we embrace and welcome new learnings derived from the data-driven approaches in education, Sentiment Analysis stands out as a stellar catalyst for all positive changes, contributing to develop a dynamic and emotionally super intelligent educational ecosystem that prioritizes the holistic and comprehensive well-being of its participants as well as all other stakeholders.

References

1. Paul LA, Quiggin J (2020) Transformative education. Educ Theory 70(5):561–579. https://doi.org/10.1111/EDTH.12444
2. Medhat W, Hassan A, Korashy H (2014) Sentiment analysis algorithms and applications: a survey. Ain Shams Eng J. Accessed 28 Jan 2024. [Online]. Available: https://www.sciencedirect.com/science/article/pii/S2090447914000550
3. Lee H, Hwang Y (2022) Technology-enhanced education through VR-making and metaverse-linking to foster teacher readiness and sustainable learning. Sustainability. Accessed 28 Jan 2024. [Online]. Available: https://www.mdpi.com/2071-1050/14/8/4786
4. Zhou J, Ye JM (2023) Sentiment analysis in education research: a review of journal publications. Interact Learn Environ. https://doi.org/10.1080/10494820.2020.1826985
5. Altrabsheh N, Gaber MM, Cocea M (2013) SA-E: sentiment analysis for education. Front Artif Intell Appl 255:353–362. https://doi.org/10.3233/978-1-61499-264-6-353
6. Lee H, Hwang Y (2022) Technology-enhanced education through VR-making and metaverse-linking to foster teacher readiness and sustainable learning. Sustainability 14. https://doi.org/10.3390/su14084786
7. Sosun SD et al (2022) Deep sentiment analysis with data augmentation in distance education during the pandemic. In: Proceedings—2022 innovations in intelligent systems and applications conference, ASYU 2022. Institute of Electrical and Electronics Engineers Inc. https://doi.org/10.1109/ASYU56188.2022.9925379
8. Shaik T et al (2022) A review of the trends and challenges in adopting natural language processing methods for education feedback analysis. IEEE Access 10:56720–56739. https://doi.org/10.1109/ACCESS.2022.3177752
9. Malviya S, Tiwari A, Srivastava R (2020) Machine learning techniques for sentiment analysis: a review. SAMRIDDHI J Phys Sci Eng Technol 12(2):72–78. https://doi.org/10.18090/samriddhi.v12i02.3
10. Hajrizi R, Nuçi KP (2020) Aspect-based sentiment analysis in education domain, Oct 2020. [Online]. Available: http://arxiv.org/abs/2010.01429
11. Zhang W, Li X, Deng Y, Bing L, Lam W (2023) A survey on aspect-based sentiment analysis: tasks, methods, and challenges. IEEE Trans Knowl Data Eng 35(11):11019–11038. https://doi.org/10.1109/TKDE.2022.3230975
12. Zhang L, Wang S, Liu B (2018) Deep learning for sentiment analysis: a survey. Wiley Interdiscip Rev Data Min Knowl Discov 8(4). https://doi.org/10.1002/WIDM.1253

13. Joseph J, Vineetha S, Sobhana NV (2022) A survey on deep learning based sentiment analysis. Mater Today Proc 58:456–460. https://doi.org/10.1016/J.MATPR.2022.02.483
14. Ravi M, Johnson SJ. Analysis of student feedback on faculty teaching using sentiment analysis and NLP techniques
15. Palli AS et al (2014) A study of sentiment and trend analysis techniques for social media content. Int J Mod Educ Comput Sci 12:47–54. https://doi.org/10.5815/ijmecs.2014.12.07
16. Liang W et al (2022) Advances, challenges and opportunities in creating data for trustworthy AI. Nat Mach Intell 4(8):669–677. https://doi.org/10.1038/s42256-022-00516-1
17. Abdullah T, Ahmet A (2022) Deep learning in sentiment analysis: recent architectures. ACM Comput Surv 55(8). https://doi.org/10.1145/3548772
18. Lou C, Atoui MA, Li X (2023) Recent deep learning models for diagnosis and health monitoring: a review of research works and future challenges. https://doi.org/10.1177/014233122 31157118
19. Sweta S, Sweta S (2021) Educational Data Mining Techniques with Modern Approach, Springer, pp. 25–38. https://doi.org/10.1007/978-981-33-4681-9_3

Chapter 4
Sentiment Tech: Exploring the Tools Shaping Emotional Analysis

4.1 Introduction

Sentiment Analysis involves the use of various latest technologies to analyse and interpret sentiments in textual/audio/video data. Machine Learning algorithms like Support Vector Machines and Deep Learning models like recurrent neural networks and BERT play a central role in automated sentiment classification. These models are envisioned on labeled datasets to recognize patterns and relationships between words and sentiments. Lexicon-based approaches, using sentiment dictionaries, offer additional context to revamp final outcomes. Natural Language Processing (NLP) libraries like NLTK and spaCy provide tools for text preprocessing and generating featured extractions. Sentiment Analysis technologies are extensively used in applications ranging from social media sentiment and other forums or chat rooms tracking to learner's feedback analysis and finding patterns from data analysis. The continuous evolution of NLP technologies contributes to the further refinement and enhancement of Sentiment Analysis's authenticity and accuracy. Emerging technologies, like transfer learning and contextual embeddings, are pushing the boundaries of Sentiment Analysis capabilities further. As Sentiment Analysis becomes more sophisticated, the integration of diverse technologies ensures a comprehensive understanding of the emotions expressed in the textual/audio/video data.

4.2 Tools and Techniques Used for Sentiment Analysis [1, 2]

From simple rule-based approaches to complex Machine Learning models and Deep Learning models, Sentiment Analysis uses a variety tools and techniques.

Figure 4.1 shows the different approaches for sentiment analysis [3–7].

Tools and techniques typically used for Sentiment Analysis are listed here.

© The Author(s), under exclusive license to Springer Nature Singapore Pte Ltd. 2024
S. Sweta, *Sentiment Analysis and its Application in Educational Data Mining*,
SpringerBriefs in Computational Intelligence,
https://doi.org/10.1007/978-981-97-2474-1_4

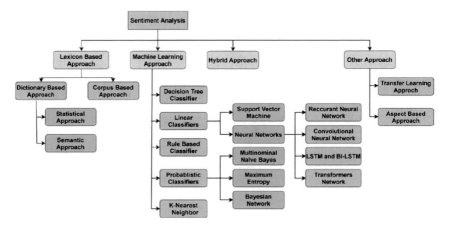

Fig. 4.1 Approaches to Sentiment Analysis

4.2.1 Lexicon-Based Approaches [8]

Lexicon-based methods based on predefined dictionaries or lexicons that attach words with sentiment scores. These methods operate on rules and do not need training on annotated datasets. Lexicon-based methods provide as a strong foundation for Sentiment Analysis, especially in situations defined by limited training data or when interpretability assumes critical importance. Machine Learning-based techniques may be more preferred, especially when dealing with diverse verbal expressions, due to their ability to provide more complex analyses and excellent performance. It is possible to create or extend sentiment lexicons to incorporate domain-specific term or to modify sentiment ratings according to settings.

Here some of the tools, techniques, method, and algorithm are mentioned which support in Lexicon-Based Approaches

- Tools:
  - NLTK (Natural Language Toolkit): A powerful Python library design for manipulating human language data. It includes a Sentiment Analysis module.
  - Text Blob: A Python library built on top of NLTK framework which simplifies text processing and offer a straightforward interface for doing Sentiment Analysis.

- Techniques:
  - Sentiment Lexicons: Lexicon-based methods use predefined word lists with associated sentiment scores to determine the sentiment of a text.
  - Rule-Based Analysis: Implement rule-based analysis and set rules to determine sentiment based on the presence of specific words or patterns in the text.

- Method:

- Text Preprocessing: Text preprocessing involves the cleaning and preprocess of the input text, which may include actions such as converting text to lowercase, eliminating stop words, and punctuation.
- Lexicon Creation: Creating a sentiment lexicon with words that are associated with sentiment scores. A score is assigned to each word, which signifies whether it conveys a positive, negative, or neutral sentiment.
- Sentiment Score Calculation: The sentiment scores for the words in the pre-processed text are computed by employing the lexicon. The aggregate sentiment score of specific words is used to determine the overall sentiment score of the text.
- Sentiment Classification: In classifying the text as Positive, Negative, or Neutral, a threshold is applied to the overall sentiment score.

- Algorithm: Lexicon-Based Sentiment Analysis

  - Input: Text (pre-processed)
  - Output: Sentiment (Positive, Negative, or Neutral)
  - Steps:
    1. Load a sentiment lexicon with words and associated sentiment scores.
    2. Initialize variables for positive_score, negative_score, and overall_sentiment_score.
    3. Tokenize the pre-processed text into words.
    4. For each word in the text:
       a. If the word is in the lexicon:

          - Add the sentiment score of the word to positive_score or negative_score.

       b. If the word is not in the lexicon:

          - Ignore the word or use a default score.

    5. Calculate overall_sentiment_score = positive_score − negative_score.
    6. Apply a threshold to overall_sentiment_score:

       - If overall_sentiment_score > 0, classify as Positive.
       - If overall_sentiment_score < 0, classify as Negative.
       - If overall_sentiment_score ≈ 0, classify as Neutral.

    7. Return the classified sentiment.
- Advantages of Lexicon-Based Approaches:

  - Simplicity and transparency.
  - No need for large labeled datasets for training.

- Limitations of Lexicon-Based Approaches:

  - Limited coverage for domain-specific terms.
  - Sarcasm, irony, or nuanced language may create a big challenge.

4.2.2 Machine Learning-Based Approaches [9]

Machine Learning-based Sentiment Analysis makes a great deal of approaches that can acquire knowledge of complex relationships and patterns within data, thereby facilitating accurate generalization to new instances. They have demonstrated effectiveness across a range of applications like Sentiment Analysis in social media, ratings from customers, and learner feedback and discussions in an academic setting.

Applying Machine Learning approaches in Sentiment Analysis focuses to using Machine Learning techniques to automatically detect and categorize sentiments in textual data. Instead of depending on humanly created rules or dictionaries, these methods apply computational models that acquire knowledge from labeled data to make predictions about the sentiment conveyed in text.

- Tools:

  - scikit-learn: It is a robust Machine Learning library for Python that includes tools for text processing and Sentiment Analysis.
  - TensorFlow and PyTorch: These are the Deep Learning frameworks that can be used for building and training complex Sentiment Analysis models.

- Techniques:

  - Naive Bayes Classifiers: Simple effective and useful for text classification, including Sentiment Analysis.
  - Support Vector Machines (SVM): Effective for binary and multiclass classification tasks, including Sentiment Analysis.
  - Recurrent Neural Networks (RNNs) and Long Short-Term Memory Networks (LSTMs): It is a Deep Learning architecture that can capture sequential information in textual data.

- Method:

  - Data Collection: Collect a labeled dataset with few examples of text and their corresponding sentiment labels (positive, negative or neutral).
  - Data Preprocessing: Clean and preprocess the text data by removing stop words, punctuation and irrelevant symbols by applying pre-processing task. Tokenize the text into tokens. Convert the text to a vector representation using techniques like TF-IDF (Term Frequency-Inverse Document Frequency) and many more available power full techniques.
  - Model Training: Split the dataset into training and testing sets. Choose an appropriate Machine Learning algorithm (e.g., SVM) and train the model on the training set.
  - Model Evaluation: Evaluate the trained model on the testing set to analyse its performance. Metrics like accuracy, precision, recall, and F1 score can be used to evaluate the model's effectiveness.
  - Prediction: Use the trained model to predict the sentiment of new, unseen text.

- Algorithm (Support Vector Machines—SVM):
  Support Vector Machines are widely used for text classification tasks specially in Sentiment Analysis due to their ability to handle high-dimensional data and non-linear association.
  Steps for SVM for Sentiment Analysis:

  - Data Representation: Represent each document as a vector in a high-dimensional space, where each dimension corresponds to a unique word or feature.
  - Feature Extraction: Use powerful techniques like TF-IDF, Word2vec or Glove to extract features from the text, assigning weights to each term based on its importance in the document.
  - Training: Train the SVM model using the labeled training dataset. The goal is to find a hyperplane that best separates the feature vectors of positive and negative examples.
  - Prediction: Given a new document, convert it into the same feature space used.

4.2.3   Deep Learning-Based Approach [10]

Deep Learning has demonstrated remarkable success in various Natural Language Processing (NLP) tasks, including Sentiment Analysis. The Deep Learning models for Sentiment Analysis leverage Neural Networks, especially recurrent Neural Networks (RNNs), Long Short-Term Memory Networks (LSTMs), and more recently, Transformer-Based Models like BERT (Bidirectional Encoder Representations from Transformers).

Deep Learning-based Sentiment Analysis approaches comprise a variety of tools, methods, techniques, and algorithms that are used to build and train complex Neural Network models.

A comprehensive analysis of each component follows:

- Tools:

  - TensorFlow: An open-source Deep Learning framework developed by Google. TensorFlow exhibits a comprehensive and holistic set of tools for building and training Deep Neural Networks. TensorFlow is generally used for implementing various Deep Learning architectures, including Recurrent Neural Networks (RNNs), Long Short-Term Memory Networks (LSTMs), and Transformer-Based Models.
  - PyTorch: An open-source Deep Learning library known for its dynamic computation graph, making it flexible for experimentation and research. PyTorch is popular for building and training Deep Learning models, particularly in research settings. It supports the implementation of various neural network architectures for Sentiment Analysis.
  - Keras: It is a high-level Neural Networks API written in Python that runs on top of TensorFlow or other backend engines. Keras simplifies the

process of building and training Deep Learning models, making it accessible for practitioners. It's often used for quick prototyping of Neural Network architectures.

- Techniques:

  - Word Embeddings: Key representing words as dense vectors in a continuous vector space. Word embeddings capture semantic relationships between words. Pre-trained word embeddings (e.g. Word2Vec, GloVe) or trainable embeddings are used to convert words into numerical representations, which are then fed into the Neural Network. The digitization of the words helps to make collate data faster and again the outcomes can be converted into words for better understanding in terms of context.
  - Recurrent Neural Networks (RNNs): Neural networks designed to handle sequences of data by introducing a loop to process information over specified timeline. RNNs are suitable for tasks where the order of words matters. Here, capturing dependencies and patterns in sequences are carried out in text.
  - Long Short-Term Memory Networks (LSTMs): A type of RNN designed to address the issue of vanishing gradient, allowing the network to capture long-term dependencies in sequences. LSTMs are effective in understanding context and dependencies in sentences in sequential format, making them suitable for Sentiment Analysis.
  - Transformer-Based Models (e.g., BERT): Attention-based models that consider the entire context of a sequence simultaneously, avoiding sequential processing limitations. Transformer-based models, such as BERT (Bidirectional Encoder Representations from Transformers), have achieved state-of-the-art results in various natural language processing tasks, including Sentiment Analysis.

- Method:

  - Data Preparation: Collect and preprocess the dataset, including task like cleaning, tokenization, and converting text data into a format appropriate for Deep Learning models.
  - Embedding Layer: Use an embedding layer to convert words into dense vectors. This layer learns word representations based on the context in which words appear.
  - Model Architecture: Design an architecture of Deep Learning model.

    Common Choices Include RNNs: Suitable for sequential data, but they may suffer from vanishing gradient problems.
    Long Short-Term Memory Networks (LSTMs): A type of RNN designed to capture long-range dependencies in sequential data.
    Transformer-Based Models (e.g. BERT): Bidirectional models that consider the context of each word in both directions, provides highly effective results.

- Training: Train the model using a labeled dataset. While adjusting hyperpa-rameters such as batch size and learning rate, observe the model's performance on validation data.
- Evaluation: Evaluate the trained model on a separate test dataset to evaluate its generalization performance.

• Algorithm:

1. Initialize Model:

   - Choose a Deep Learning architecture (e.g. LSTM, BERT) for Sentiment Analysis.
   - Initialize the model with the embedding layer.

2. Embedding Layer:

   - If pre-trained word embeddings are available, load them into the embedding layer.
   - If not, let the model learn word embeddings during training.

3. Model Architecture:

   - Define the architecture of the Deep Learning model, including the choice of layers, output layer and activation functions.

4. Compile Model:

   - Compile the model by specifying the loss function, evaluation metric and optimizer.

5. Train Model:

   - Train the model on the labeled dataset.
   - Adjust hyperparameters based on the performance on validation data.

6. Evaluate Model:

   - Evaluate the trained model on a separate test dataset to evaluate its accuracy and generalization.

7. Predict Sentiments:

   - Use the trained model to predict sentiments for new, unseen data.

• Example (using Keras with LSTM): Implementation

```
# Import required library
from keras.models import Sequential
from keras.layers import Embedding, LSTM, Dense
# Assuming 'X_train', 'y_train' are your training data and labels
# Define the model
model = Sequential()
model.add(Embedding(input_dim = vocab_size, output_dim = embedding_dim,
input_length = max_len))
```

```
model.add(LSTM(units = 100))
model.add(Dense(units = 1, activation = 'sigmoid'))
# Compile the model
model.compile(optimizer = 'adam', loss = 'binary_crossentropy', metrics =
['accuracy'])
# Train the model
model.fit(X_train, y_train, epochs = 5, batch_size = 32, validation_split = 0.1)
# Evaluate the model
loss, accuracy = model.evaluate(X_test, y_test)
print(f'Accuracy: {accuracy}')
# Make predictions
predictions = model.predict(new_data)
Note:
```

- This example uses a simple LSTM-based architecture. BERT or other transformer-based models would require a different set of libraries and implementation.
- vocab_size, max_len, embedding_dim, X_train, X_test, y_train, and y_test need to be appropriately defined based on your dataset.
- This example demonstrates a basic Sentiment Analysis model implemented using an LSTM architecture in Keras. Depending on the dataset and requirements, we may choose more powerful architectures or pre-trained models for improved performance.

These tools, techniques, methods, processes, and algorithms collectively form the foundation of Deep Learning-Based Approaches in Sentiment Analysis. Researchers and practitioners select and customize these components based on the specific requirements of the Sentiment Analysis task and the characteristics of the data at hand for further refinement.

4.2.4   Hybrid Approaches [11]

Hybrid Approaches provide a flexible framework for Sentiment Analysis, allowing the required customization based on the specific requirements and challenges of different applications. It helps to address critical aspects of both approaches which are not possible to address individually and simultaneously. The choice of integration strategy and combination of methods depends on the characteristics of the data, datasets and the goals of the Sentiment Analysis tasks. Hybrid Approaches in Sentiment Analysis combine multiple methods, models, scenario or features to enhance the overall performance and robustness of Sentiment Analysis systems. These approaches aim to leverage the strengths of different techniques and processes to address the limitations of individual methods for best outcome.

Hybrid approaches in Sentiment Analysis combine the strengths of lexicon-based methods and Machine Learning (ML) techniques to achieve more accurate and comprehensive sentiment predictions.

Here's an overview of key components and the characteristics of Hybrid Approaches are discussed as under:

4.2.4.1 Key Components

- Lexicon-Based Features: Lexicon-based Sentiment Analysis depends on predefined dictionaries of words and their linked sentiment scores.

  – Integration: Lexicon-based features can be combined with other approaches to provide a basic sentiment score or as additional features for Machine Learning Models.

- Machine Learning Models: Supervised Machine Learning Models like Support Vector Machines (SVM), Naive Bayes, or Random Forest, are trained on labeled data to predict sentiments in more precise way.

  – Integration: The predictions from Machine Learning models can be combined with other methods for a more comprehensive and value additive Sentiment Analysis.

- Deep Learning Models: Deep Learning Models, like Recurrent Neural Networks (RNNs) or Transformer-Based Models (e.g., BERT), capture complex relationships in text.

  – Integration: Deep Learning Models can be combined with other methods to benefit from their ability to understand the contextual distinctive aspects.

- Aspect-Based Analysis: Instead of analysing sentiment at the document level, aspect-based analysis breaks down the sentiments into finer categories related to specific aspects or topics within the text and enable to provide more insights.

  – Integration: Aspect-based Sentiment Analysis can be integrated with other methods to provide more detailed insights, especially in the contexts with multiple aspects as discussed above.

- Rule-Based Systems: Rule-based systems use predefined rules to identify sentiments based on linguistic features, patterns, processes, or keywords.

  – Integration: Rule-based systems can complement other methods, especially in cases where specific sentiment expressions can be clearly defined and linked to variables to get better results.

4.2.4.2 Integration Strategies

- Ensemble Methods: Ensemble methods combine and consolidate predictions from multiple models to improve overall accuracy, efficiency and robustness of the system. It helps to accumulate benefits of all the models.
  Example: Combining analyses and predictions from a lexicon-based approach, a Machine Learning model and a Deep Learning model to obtain finally improved sentiment score. It also helps to compare different scores obtained at different point of time and provides development plan for further improvement in score.
- Sequential Processing: Applying different methods sequentially, where the output of one method serves as input or features for another. The outcomes can be studied by changing order of the sequence of learning processes to find out the best sequence and its predictions.
  Example: Using lexicon-based analysis to pre-process data and then feeding the outcomes into a Machine Learning model for further refinement and improvement.
- Weighted Combination: Assigning different weights to the outputs of individual methods based on their reliability or performance.
  Example: Giving higher weightage to the predictions of a Machine Learning model if it has demonstrated better accuracy and efficiency parameters during evaluation process.
- Contextual Switching: Dynamically selecting the most suitable method based on the context or characteristics of the input data. It is most important part of the modelling to select the best model to improve performance while the datasets change dynamically.
  Example: Choosing between lexicon-based analysis and Machine Learning models based on the length, quality or complexity of the text.

4.2.4.3 Advantages of Hybrid Approaches

- Improved Accuracy: By combining the positive attributes of different methods, hybrid approaches can often achieve higher accuracy and efficiency compared to individual methods.
- Robustness: Hybrid approaches are more robust to variations in data and can handle diverse linguistic patterns comfortably.
- Adaptability: They can adapt to different types of text and contexts, making them versatile in terms of various Sentiment Analysis applications.
- Addressing Data Limitations: In the given scenarios with limited labeled data, hybrid approaches can be leveraged through unsupervised methods or pre-trained models.

4.2.4.4 Challenges Hybrid Approaches

- Complexity: Designing and implementing hybrid approaches can be more complex, cumbersome, and difficult to get desired information than the individual methods.
- Interpretability: Combining multiple methods may provide outcomes challenging to interpret with the reasons behind the specific patterns and predictions.
- Tools:
  - VADER (Valence Aware Dictionary and Sentiment Reasoner): A rule-based Sentiment Analysis tool designed for social media text that combines Machine Learning approaches and lexicon-based approaches.
- Techniques:
  - Combining ML and Lexicon Methods: Hybrid models improved accuracy by focusing on the positive connection of lexicon-based and Machine Learning approaches.
- Method:
  - Lexicon-Based Preprocessing: Begin with lexicon-based preprocessing to extract sentiment-related features from the text. This involves using sentiment lexicons to associate sentiment scores to words or phrases in the text.
  - Feature Extraction: Utilize the sentiment scores obtained from lexicon-based preprocessing as features for ML models. These features can be combined with other traditional textual features such as word embeddings or TF-IDF (Term Frequency-Inverse Document Frequency).
  - Machine Learning Model Training: Train a Machine Learning model (e.g., Naive Bayes, SVM, or a Neural Network) using the combined features from both lexicon-based preprocessing and traditional textual features. This step involves using a labeled dataset with known sentiments.
  - Fine-Tuning: The ML model should be fine-tuned to precisely adapt its properties to the Sentiment Analysis task at hand. This may involve adjusting hyper-parameters, selecting the appropriate model architecture, or using techniques like cross-validation.
  - Integration: Combine the predictions from the lexicon-based approach and the ML model. The integration can be achieved through the implementation of more complex algorithms that account for the confidence of each approach or by calculating the weighted average of the predictions.
- Algorithm:
  1. Lexicon-Based Preprocessing:
     a. Input: Textual data
     b. Output: Sentiment scores for words/phrases
  2. Feature Extraction:
     a. Input: Sentiment scores, additional textual features

     b. Output: Combined feature set for ML model
3. Training a Machine Learning Model:
     a. Input: Combined feature set, labeled dataset
     b. Output: Trained ML model
4. Fine-Tuning:
     a. Input: Trained ML model, validation dataset
     b. Output: Fine-tuned ML model
5. Integration:
     a. Input: Lexicon-based sentiment predictions, ML model predictions
     b. Output: Integrated sentiment predictions.

4.2.5 Pre-trained Models [12]

Sentiment Analysis using pre-trained models involves empowering models that have been trained on large datasets to understand the sentiment expressed in text.

Here's an overview of the tool, techniques, method, algorithm, and an example based on pre-trained models:

- Tools:

  - BERT (Bidirectional Encoder Representations from Transformers): A pre-trained transformer-based model often fine-tuned for Sentiment Analysis tasks.
  - GPT (Generative Pre-trained Transformer): Another transformer-based model that can be fine-tuned for Sentiment Analysis.

- Techniques:

  - Transfer Learning: Leveraging pre-trained language models and fine-tuning them on specific Sentiment Analysis tasks.

- Method:

  - Choose a Pre-trained Model: Select a pre-trained Sentiment Analysis model. Common choices include BERT, GPT, RoBERTa, and others. These models are often pre-trained on massive corpora and fine-tuned for Sentiment Analysis tasks.
  - Fine-Tuning (Optional): Fine-tuning is optional but can enhance performance on specific tasks. Fine-tuning involves training the pre-trained model on a smaller dataset specific to your Sentiment Analysis task.
  - Tokenization: Tokenize the input text into smaller units (tokens) that the model can process. Many pre-trained models use subword tokenization for better handling of rare words.
  - Inference: Pass the tokenized text through the pre-trained model for inference. The model outputs a probability distribution over different sentiment class (positive, negative or neutral).

  – Post-processing: Apply post-processing to interpret the model's output. This may involve thresholding the probabilities to determine the final sentiment class.

- Algorithm:

  1. Load Pre-trained Model: Load the pre-trained Sentiment Analysis model into memory. This can be done using the specific library or framework associated with the chosen model (e.g. transformers for BERT or Hugging Face's Transformers library).
  2. Tokenization: Tokenize the input text using the model's tokenizer. This step converts the text into a format suitable for the model's input.
  3. Inference: Pass the tokenized input through the pre-trained model to obtain predictions. The model outputs a probability distribution over possible sentiment class.
  4. Post-processing: Apply post-processing to interpret the model's output. This may involve selecting the sentiment class with the highest probability or applying a threshold for binary classification.

- Example (Using Hugging Face's Transformers with BERT):
  In this example, the Sentiment Analysis model is loaded from the Hugging Face's Transformers library, tokenization is performed using the BERT tokenizer, and the model is used for inference. The post-processing step involves selecting the sentiment class with the highest probability.
  Note: Fine-tuning is often done on a dataset specific to your task, but for clarity, this example uses the pre-trained model without fine-tuning.

This approach allows you to leverage the knowledge encoded in pre-trained models, making it particularly useful when you have limited labeled data for your specific Sentiment Analysis task.

4.2.6 Cloud-Based Services [13]

Cloud-based services offer convenient and scalable solutions for Sentiment Analysis, allowing users to robust powerful models without the need for extensive infrastructure.

Here's a guide on the tools, techniques, method, algorithm, and an example based on Cloud-Based Services Approaches:

- Tools:

  – Cloud-based services such as Amazon Comprehend, Google Cloud Natural Language API, and Azure Text Analytics offer Sentiment Analysis in addition to an extensive range of functionalities for natural language processing.

- Techniques:

- API Integration: Use the Sentiment Analysis capabilities provided by cloud services through APIs for quick and scalable analysis.

- Method:
  - Choose a Cloud Service Provider: Select a cloud service provider that offers Sentiment Analysis capabilities as part of its natural language processing (NLP) services. Common providers include Amazon Web Services (AWS), Microsoft Azure, and Google Cloud Platform (GCP).
  - Set Up an Account and Access the NLP Service: Create an account with the chosen cloud service provider and set up access to their NLP services. This often involves obtaining API keys or credentials.
  - Access the Sentiment Analysis API: Utilize the API provided by the supplier to have access to the Sentiment Analysis capability. Typically, this involves making HTTP requests to the Sentiment Analysis endpoint.
  - Submit Text Data for Analysis: Send the text data (such as reviews, comments, or other textual content) to the Sentiment Analysis API for analysis. The API will return the sentiment polarity (positive, negative, or neutral) along with a confidence score.
  - Interpret Results: Interpret the results returned by the API, considering both the sentiment polarity and the confidence score. Higher confidence scores typically indicate more reliable predictions.

- Algorithm:
  The specific algorithms used by cloud-based Sentiment Analysis services are exclusive and not always disclosed. However, these services generally employ a combination of Machine Learning models, Sentiment Lexicons, and Natural Language Processing techniques. The models are often trained on large datasets to identify patterns and sentiments in text.

- Example (Using Google Cloud Natural Language API):
  Using cloud-based services simplifies the implementation of Sentiment Analysis, making it accessible to developers without the need for extensive Machine Learning expertise. It is important to consult the documentation provided by the selected cloud provider to obtain precise information and guidelines on how to utilize the API.

4.2.7 Open-Source Sentiment Analysis Frameworks [13]

Flair is an open-source NLP library in Python that provides various functionalities for various NLP tasks, including Sentiment Analysis.

- Tools:
  - Stanford NLP: A Natural Language Processing library that includes tools for Sentiment Analysis.

- Flair: An open-source framework with many functionalities related to NLP tasks, including Sentiment Analysis.

- Techniques:

  - Flair implements Sentiment Analysis using deep, layered architectures, bidirectional language models, and contextual embeddings. These techniques empower Flair to understand the complexities of language and context, hence proving its effectiveness in Sentiment Analysis task across diverse applications. Method: Sentiment Analysis using Flair Framework.

- Method:

  - Contextual Embeddings: Contextual embeddings allow Flair to understand the changing semantics of words based on their positions in a sentence, enhancing the model's ability to figure out sentiment in complex language.
  - Bidirectional Language Models: Bidirectional models enable Flair to consider both preceding and following words while making predictions. This bidirectional context is complex for understanding the full meaning of a word or phrase, especially in the context of Sentiment Analysis.
  - Stacked Embeddings and Deep Architectures: Stacking embeddings and using deep architectures enables Flair models to capture complex relationships within sentences. This depth in the architecture helps in learning hierarchical representations of language, allowing for a more detailed understanding of sentiment.
  - Embedding Techniques: By providing a range of embedding options, Flair allows users to choose the most appropriate representation for their specific Sentiment Analysis task, which offer more flexibility and adaptability.
  - Training on Large Datasets: Pre-training on complex datasets helps Flair models capture common language patterns, ambiguity, and sentiment contexts. This pre-training contributes to the model's adaptability to a wide range of domains and languages.
  - Fine-Tuning and Transfer Learning: Fine-tuning enables the customization of pre-trained models, making them more effective for specialized Sentiment Analysis applications without requiring complex training from scratch.
  - Integration with Flair NLP Library: Integration with the Flair library allows users to benefit from a unified and consistent platform for multiple NLP tasks, streamlining workflows and facilitating ease of use.

- Implementation:

  1. Install Flair:
     pip install flair
  2. Load Flair's Sentiment Analysis Model:
     from flair.models import TextClassifier
     from flair.data import Sentence
     # Load pre-trained Sentiment Analysis model
     classifier = TextClassifier.load('en-sentiment')

3. Analyse Sentiment of Text:

```
def analyze_sentiment(text):
    # Create a Flair Sentence
    sentence = Sentence(text)
    # Predict sentiment
    classifier.predict(sentence)
    # Get sentiment label and score
    sentiment_label = sentence.labels[0].value
    sentiment_score = sentence.labels[0].score
    return sentiment_label, sentiment_score
# Example usage
feedback_text = "I really enjoyed the workshop. The content was informative."
sentiment_label, sentiment_score = analyze_sentiment(feedback_text)
print(f"Sentiment: {sentiment_label}")
print(f"Sentiment Score: {sentiment_score}")
```

- Explanation:
  Loading the Model: The TextClassifier.load('en-sentiment') statement loads a pre-trained Sentiment Analysis model for English.
  Analysing Sentiment: The analyz_sentiment function takes a text input, creates a Flair Sentence, predicts the sentiment using the loaded model and retrieves the sentiment label and score.
  Output: The sentiment label is the predicted sentiment class (e.g. POSITIVE, NEGATIVE), and the sentiment score is a confidence score associated with the prediction.
- Algorithm: Flair-based Sentiment Analysis Algorithm

  – Load Pre-trained Model: Load a pre-trained Sentiment Analysis model using Flair.
  – Preprocess Text: Pre-process the text data if needed (e.g., cleaning, tokenization).
  – Create Flair Sentence: Create a Flair Sentence object using the input text.
  – Predict Sentiment: Use the loaded model to predict the sentiment of the sentence.
  – Retrieve Results: Retrieve the sentiment label and score from the model's predictions.
  – Output: Return or use the sentiment label and score as required.

Note:

- The Flair framework provides pre-trained models for Sentiment Analysis in multiple languages. You can choose the appropriate model based on your language needs.
- This example uses a simple function for analysis. For more complex applications, you can incorporate Sentiment Analysis into a larger NLP pipeline using Flair.

- Customization: Some frameworks allow customization and fine-tuning for specific Sentiment Analysis tasks.
- These tools and techniques can be chosen based on factors such as the complexity of the Sentiment Analysis task, the size of the data set and the available resources for training and implementation. The choice often depends on the specific requirements and constraints of the project at hand.

4.3 Conclusion

In conclusion, the dynamic landscape of Sentiment Analysis is intricately shaped by a diverse array of tools and techniques. From traditional lexicon-based methods to sophisticated Machine Learning and Deep Learning approaches, each tool and technique contributes a unique set of strengths to the realm of Sentiment Analysis. The efficiency of Sentiment Analysis greatly benefits from Natural Language Processing (NLP) libraries like NLTK, spaCy, and Flair, which provide robust tools for text preprocessing and feature extraction. As the field evolves, the integration of advanced technologies, including transfer learning and contextual embeddings, showcases a commitment to enhancing accuracy and adaptability. The amalgamation of these tools not only enables us to comprehend sentiments across vast datasets but also empowers applications in diverse domains, from Business Intelligence to Educational Data Mining. Moving forward, the continued exploration and integration of emerging tools and techniques will undoubtedly propel Sentiment Analysis into new realm of sophistication, refining our ability to decipher the emotion of human expression in textual data.

References

1. Bairam M, Abdullah A, Aqlan Q, Manjula B, Lakshman Naik R (2019) A study of sentiment analysis: concepts, techniques, and challenges, vol 28, pp 147–162. Springer. https://doi.org/10.1007/978-981-13-6459-4_16
2. Dolianiti F, Iakovakis D, Dias SB et al (2018) Sentiment analysis techniques and applications in education: a survey. In: International conference on technology and innovation in learning, teaching and education. Springer. Accessed 28 Jan 2024. [Online]. Available: https://doi.org/10.1007/978-3-030-20954-4_31
3. Shaik T et al (2022) A review of the trends and challenges in adopting natural language processing methods for education feedback analysis. IEEE Access 10:56720–56739. https://doi.org/10.1109/ACCESS.2022.3177752
4. Zhang L, Wang S, Liu B (2018) Deep learning for sentiment analysis: a survey. Wiley Interdiscip Rev Data Min Knowl Discov 8(4). https://doi.org/10.1002/WIDM.1253
5. Medhat W, Hassan A, Korashy H (2014) Sentiment analysis algorithms and applications: a survey. Ain Shams Eng J. Accessed 28 Jan 2024. [Online]. Available: https://www.sciencedirect.com/science/article/pii/S2090447914000550
6. Joseph J, Vineetha S, Sobhana NV (2022) A survey on deep learning based sentiment analysis. Mater Today Proc 58:456–460. https://doi.org/10.1016/J.MATPR.2022.02.483

7. Wankhade M, Rao ACS, Kulkarni C (2022) A survey on sentiment analysis methods, applications, and challenges. Artif Intell Rev 55(7):5731–5780. https://doi.org/10.1007/s10462-022-10144-1
8. Ahmad M, Aftab S, Muhammad S, Waheed U (2017) Tools and techniques for lexicon driven sentiment analysis: a review. Int J Multidiscip Sci Eng. Accessed 28 Jan 2024. [Online]. Available: http://www.ijmse.org/Volume8/Issue1/paper4.pdf
9. Malviya S, Tiwari A, Srivastava R (2020) Machine learning techniques for sentiment analysis: a review. SAMRIDDHI J Phys Sci Eng Technol 12(2):72–78. https://doi.org/10.18090/samriddhi.v12i02.3
10. Abdullah T, Ahmet A (2022) Deep learning in sentiment analysis: recent architectures. ACM Comput Surv 55(8). https://doi.org/10.1145/3548772
11. Ahmad M, Aftab S, Ali I, Hameed N (2017) Hybrid tools and techniques for sentiment analysis: a review. Int J Multidiscip Sci Eng 12(4):2021. Accessed 28 Jan 2024. [Online]. Available: https://www.academia.edu/download/53953860/paper5.pdf
12. Cambria E, Das D, Bandyopadhyay S, Feraco A (2017) A practical guide to sentiment analysis. https://doi.org/10.1088/1757-899X/551/1/012070
13. Singh LG, Singh SR (2021) Empirical study of sentiment analysis tools and techniques on societal topics. J Intell Inf Syst 56(2):379–407. https://doi.org/10.1007/S10844-020-00616-7

Chapter 5
Emerging Trends and Challenges in Educational Sentiment Analysis

5.1 Introduction

Sentiment Analysis in Educational Data Mining has opened many potential revenues. It has completely revolutionized educational domain in view of important in-sites and scientific patterns immerging out of data analysis. It has huge potential to elaborate contemporary educational trends and it will provide knowledge regarding evolving inputs, promising opportunities.

With help of Sentiment Analysis, several challenges of EDM can be addressed and full potential of such analysis can be exploited for the benefits of all stakeholders [1, 2].

In this chapter, we explore the literature review, emerging trends and innovations along with the challenges in educational Sentiment Analysis which highlighting the emerging trends and further shaping the future of this evolving field.

5.2 Literature Review [2]

Some literature reviews related to Sentiment Analysis in EDM have been explained as under Table 5.1.

5.3 Emerging Trends in Educational Sentiment Analysis [3]

Sentiment Analysis has taken as centre state in EDM in recent past. This area of studies has witnessed tremendous growth and created a promising research field among various researcher and academicians. It holds future promises to redefine

© The Author(s), under exclusive license to Springer Nature Singapore Pte Ltd. 2024
S. Sweta, *Sentiment Analysis and its Application in Educational Data Mining*,
SpringerBriefs in Computational Intelligence,
https://doi.org/10.1007/978-981-97-2474-1_5

Table 5.1 Sentiment Analysis on education

Educational application	Algorithm used	References
Learning and teaching systems Evaluation	Ensemble learning Association rule mining RF word2vec, fastText, GloVe, LSTM	Lalata et al. (2019) Roaring et al. (2022) Pramod et al. (2022) Onan (2020)
Decision-making	VSF, VGFSS, SVM, ANN	Hussain et al. (2022)
Enhance pedagogical concepts	LDA, LIWC Topic ontology Reinforcement learning	Yan et al. (2019) Li et al. (2022a) Feidakis et al. (2019)
Assessment evaluation	Vader Additive regression RF, SVN, KNN	Janda et al. (2019) Gkontzis et al. (2020)

educational covenants. There is great hope among all stakeholder to increase the effectiveness of educational systems.

Some important emerging trends are mentioned in Table 5.2 [3].

5.3.1   Emerging Tools Used in Educational Sentiment Analysis

- BERT (Bidirectional Encoder Representations from Transformers):
  Description: A pre-trained transformer-based model for natural language understanding, widely used in Sentiment Analysis.
  Application: Fine-tuning BERT for Sentiment Analysis on educational text data.
- VADER (Valence Aware Dictionary and sEntiment Reasoner):
  Description: A lexicon and rule-based Sentiment Analysis tool that is sensitive to both polarity and intensity.
  Application: Analysing sentiment in educational social media discussions.
- GPT (Generative Pre-trained Transformer):
  Description: Language models trained on vast amounts of data, capable of generating human-like text.
  Application: Using GPT-based models for generating educational content with a positive sentiment.
- FastText:
  Description: Library for efficient learning of word representations and text classification.
  Application: Classifying sentiments in short educational texts or comments.
- DeepMoji:
  Description: Deep Learning model for learning and predicting emoji representations from text.

Table 5.2 Technology used in Sentiment Analysis

Author(s)	Research paper title	Techniques or methods used	Key points	Output/precision
Smith et al. (2020)	"Deep Sentiment Analysis: A Comprehensive Review"	Deep Learning, LSTM, CNN, Word Embeddings	Explores various deep learning techniques for sentiment analysis, compares their performance	85.2%
Garcia et al. (2017)	"Hybrid Approach for Sentiment Analysis: Combining Rule-Based and Machine Learning"	SVM, Rule-Based, Feature Engineering	Proposes a hybrid approach combining rule-based and machine learning for improved accuracy	80.5%
Biswas Jeet et al. (2021)	Sentiment Analysis Using AI	Multinomial Naive Bayes (MNB), SVM, Hidden Markov Model (HMM), Long LSTM, BERT	Analyses the effectiveness of MNB, SVM, HMM, LSTM and BERT methods in sentiment analysis	Comparison between model accuracy
Brown et al. (2019)	"Ensemble Learning for Sentiment Analysis: A Survey"	Random Forest, Bagging, Boosting	Investigates the impact of ensemble learning on sentiment analysis	87.9%
Catelli et al. (2022)	Lexicon-Based Versus BERT-Based Sentiment Analysis: A Comparative Study in Italian	Deep Learning and BERT	Work with Italian language. BERT in the Italian language	SHAP tool was used. SHAP employs a generic approach to explicate the predictions of any given model

(continued)

Table 5.2 (continued)

Author(s)	Research paper title	Techniques or methods used	Key points	Output/precision
Kim et al. (2021)	"Transfer Learning in Sentiment Analysis: A State-of-the-Art Review"	BERT, Transfer Learning, Fine-Tuning	Reviews the application of transfer learning, specifically BERT, in sentiment analysis	92.3%
Arwa Diwali et al.	Sentiment Analysis Meets Explainable Artificial Intelligence: A Survey on Explainable Sentiment Analysis	eXplainable Artificial Intelligence (XAI) Methodologies	Lime, SHAP	A comprehensive review of sentiment analysis explainability

Application: Incorporating emoji-based Sentiment Analysis in educational chat applications.

5.3.2   Emerging Techniques Used in Educational Sentiment Analysis

- Attention Mechanisms:
  Description: Techniques that allow models to focus on specific parts of input data, enhancing Sentiment Analysis in nuanced educational contexts.
  Application: Improving Sentiment Analysis in lengthy learner essays or complex educational discussions.
- Federated Learning:
  Description: Training Machine Learning models across decentralized devices without exchanging raw data.
  Application: Collaborative Sentiment Analysis across educational institutions without compromising data privacy.
- Ethical AI in Education:
  Description: Integrating ethical considerations into Sentiment Analysis models to avoid biases and ensure fair evaluation.
  Application: Building Sentiment Analysis systems that consider cultural and demographic diversity in educational settings.

5.4 Opportunities in Educational Sentiment Analysis

Sentiment Analysis has opened a wide range of new opportunism for all stakeholders to testify the existing models and various educational plans. It will reduce biasness of the input providers as they will be crossed verified by behavioural, activities on social platforms and semantic para phrases written by the input providers. More in-depth study will provide various linkages among several variables of sentiments. The validated permutations and combination will lead to improve the efficacy of the educational system and will shape the future course of action. The derived knowledge will signify the importance of sentiments and other emotional characteristics during the pre and post learning process.

Some important evolving and potential opportunities are given in Table 5.2 to describe opportunities for Advancements in Sentiment Analysis and Educational Data Mining [4].

Educational Sentiment Analysis is a rapidly evolving field that holds great potential for transforming education. By analysing and understanding the sentiments expressed by learners, educators, and other stakeholders in the educational domain, valuable insights can be gained to enhance efficiency and effectiveness of overall teaching and learning experiences.

Sentiment Analysis and Educational Data Mining have revolutionized the field of education by providing valuable insights into learner's sentiments, emotions, and attitudes. As technology continues to evolve, there are exciting opportunities for advancements in Sentiment Analysis and Educational Data Mining. This point explores the potential areas for growth and innovation in the field, highlighting the opportunities to enhance personalized learning, implement recommender system, improvise educational outcomes by further fine tuning of the variables and shaping the future of robust educational system.

5.4.1 Fine-Grained Sentiment Analysis [5]

Leaner shows different sentiments at different points of time; however some sentiments can be measured, and some cannot because of limitation of tools capturing all sentiments. Some fine grain sentiments are deeply rooted which can be detected by capturing the exhibits of emotions, behaviour and opinions on social media platforms, performing different skills and depicting changes in attitudinal aspects and getting regulated marks (maximum or minimum) in same or different situations and mathematical applications. Such analysis will reveal personalized interactions in adaptive approaches and exchange between learner-machine on emotional and cognitive notes. There is a further scope of observing and studying other sentiments to get more fine-tuned result based on high end computational and mathematical algorithms and supporting machines. More accuracy and precision can be achieved through better interpretation of sentiments and related data analysis.

5.4.2 *Emotion Recognition and Engagement Assessment [6, 7]*

As the development of digital revolutions is growing day by day, more and more highly sophisticated and high-powered system has been developed to sense the changes in human behaviour. It helped to capture, assess, and analyse such sentiments more precisely which helping to understand the learner and requirement of all stakeholders in a seamless manner. The improvised system can integrate with facial expression and its analysis, changes in physiological changes, speech modulations, social media posts, asking questions, writing blogs, chatting in virtual rooms, etc. Understanding these attributes is very essential to ensure effective educational system. They help in getting valuable insights and resultantly tailor made educational and instructional commands can be generated. The entire process will help to develop adaptive learning, intelligent tutoring and personalized recommending system. It will also enhance the level of engagements and foster more supportive system for development of better educational environments.

5.4.3 *Cross-Domain Sentiment Analysis [8]*

Studying human behaviour and phycology is always fascinating. The sentiments may be positive or negative at one point of time and they may change at different point of time and varying situations. The sentiments are sensed across the multiple educational domains and cross-functional areas. They may exhibit different characteristics as per changes in subject, context, application or approach, etc. Different sentiments may be occurred at the time of facing questions and situations related to SWOT and trends or pressure or challenges to compete the tasks as well as exams in the stipulated timeframe in evidence-based decision-making scenario. The varying sentiments from courses, mentors, departments, social sites, and institutions may be consolidated and a comprehensive view of the learner can be identified. But the basic ingredients of the sentiments will be revealed from the analysis of information from these transcending specific courses and disciplines. They help to understand things clearly and to promote best practices among all stakeholders. The cross-domain Sentiment Analysis helps to transfer the latest knowledge and share the best adaptive model or process among all stakeholders covering all educational settings, subjects, and learners.

5.4.4 Multilingual Sentiment Analysis [9]

World is culturally diverse with so many races, ethnicities, religions, countries, languages, etc. Each one like language-specific features, cultural variations, or sentiment lexicons has dominance on behavioural and psychological aspects. Globalized world has made educational data more multilingual and multicultural. The new analysis of multilingual sentiments shows that sentiments expressed in different domains have provided comprehensive and futuristic results. Further advancement of such analysis can explore more effective way to handle cultural variations by overcoming language barriers to help such learners and empower all stakeholders.

5.4.5 Contextual Sentiment Analysis [10]

Empirical study of sentiments may provide positive, negative, or neutral sentiments. However, its true analysis will be only possible in which context it was interpreted. The emerging trends in Sentiment Analysis is to study and analyse learner's trails in view of course contents, topics, subjects, learner's profile and demographic identity, economic and social fabrics and its mutual interactions. These aspects help to understand deeper insights and lead to adaptive changes to suit all stakeholders.

5.4.6 Multimodal Sentiment Analysis [11]

Traditionally, we were doing textual data analysis including sentiments as much as they deciphered. Many new possibilities of analysis and insights of sentiments have been found with integration of several available multimedia tools, packaged software, social media sites, AI and Machine Learning tools, facial expressions, and behavioural traits. The combination of natural language processing and computer's audio/video recognition techniques expressed through verbal as well as non-verbal cues has led finding of new innovative ways to learn and understand expressed emotions quickly, smartly and comprehensively. It helps to develop more holistic and complete sentimental educational experience.

5.4.7 Real-Time Sentiment Analysis [11]

As of now, every analysis is discussed either related to online or offline. Online or real-time analysis has advantages over traditional feedback mechanism. It empowers all stakeholders including learners, administrators, and research scholars to have immediate feedback and provides opportunities to adaptively change the course of action in between. The entire processes can be done during live learning, streaming, interaction, and performance. New innovative tools in natural language processing algorithms and computerized streaming data analysis enable Sentiment Analysis to analyse live situations in real-time. It enables to adapt things dynamically and facilitate all stakeholders to do effective communication and purposeful engagements.

5.4.8 Social Network Analysis and Sentiment

As more and more people engage with social media platforms, it will be easy for academicians and research scholars to analyse the related social media sentiments easily like taking cues from online discussions, chatting, evidence from online forums and archival data from such platforms and their changing patterns over a period of time, etc. which can be utilized effectively to augment better learning for all stakeholders. It also provides the ideas of group sentiments and a bit of social and economic fabric of the learner. The integration of social network analysis and Sentiment Analysis provides a powerful mechanism to identify sentiments of the power centre of the group, leading members of the group, idea behind formation of the group, characteristics of the clusters within the group, influential person/community, and related static and dynamic matrices.

5.4.9 Explainable Sentiment Analysis [12–14]

As we know emotions play crucial roles in our day-to-day activities; therefore, the Sentiment Analysis will explain the cause-effect analogies for a particular scenario. A more granular interpretation related to explainable Sentiment Analysis provides the transparent view in the analysis process, enabling all stakeholders in educational domain to understand sentiments assigned to certain texts or contexts. So that their decision-making processes would be more precise and based on some evidential logics.

5.4.10 Domain-Specific Sentiment Lexicons [15]

Domain-specific sentiments are linked with the different educational activities. It is understood after studying and segregating emotion-action matrices. So, for further developing domain-specific sentiments lexicons are tailored to the specific educational texts, audio, video, etc. It helped to improve the accuracy and relevance of Sentiment Analysis in educational domain.

5.4.11 Integration with Educational Analytics [16]

There is a huge impact of emotions over different learning processes. Therefore, the Sentiment Analysis has edge to complement, correlate, and integrate with other educational analytical techniques like learning analytics, social network analysis, and predictive modelling. It gives a more holistic view of learner's experiences and core interactions. It enhances the predictive analysis, accuracy, and effectiveness of Sentiment Analysis, having comprehensive insights into the factors influencing learner's sentiments and improvised academic success learning outcomes can be obtained for all stakeholders.

5.4.12 Ethical Considerations and Responsible Use

Ethics has critical role in ensuring effective and efficient education systems. Ethical considerations and prevalent practices have great impact on outputs of the sentiments analysis. All stakeholders including educational institutes, main database servers, learner's profile and personalized data must ensure data protection and ethical conduct in view of handling such data. Prevalent guidelines, pre-defined protocols, and governance frameworks that outline the responsible use of Sentiment Analysis results to protect the rights and well-being of learners and all stakeholders while leveraging the benefits of Sentiment Analysis for their educational improvements and accomplishments.

5.5 Challenges in Educational Sentiment Analysis [4, 17, 18]

Getting better results to improvise educational system in view of adding dimensions of Sentiment Analysis require to select the most important variables of the sentiments. It is challenging to map, capture, and evaluate all variables in a computational model which may not be feasible. Moreover, adding more variables will give results closer to real-world problem solving. We foresee that things will improve later, and many influential factors may be added in the study to further fine tune the Sentiment Analysis and effectiveness of the education system will be upgraded accordingly.

But mapping and evaluating all factors may not be feasible in normal situations and is true that Sentiment Analysis has added many new dimensions in Educational Data Mining. However, there are still many challenges on part of capturing entire variables related to educational sentiments into computerized model for the further analysis and research.

We are hopeful that sooner or later things will improve, and many more dimensions of educational sentiments can be captured in this regard to get more fine tune results to further improve the effectiveness of education system.

Some important challenges are mentioned below.

5.5.1 Data Privacy and Ethics

It is always challenging to ensure confidentiality, privacy, safety, security, and ethical protection of data related to all stakeholders especially of the learner's while data mining. As the Sentiment Analysis involves analysing a lot of personal textual/audio/video data, protecting learner privacy and complying with data protection regulations are very critical to upheld. Implementing robust data anonymization techniques, obtaining informed consent, and establishing ethical guidelines for the Sentiment Analysis in educational domains are highly required for development of EDM.

5.5.2 Handling Subjectivity and Context

Learning epitome is achieved by holistic approach. Analysing the factors of subjectivity and contextual variations is big challenge in Sentiment Analysis. Educational texts/audio/video/or posts on social media or chat box often contain complex and domain-specific language, sarcasm, irony, personal, social, and cultural references that change sentiment interpretations as per the situations and exhibited emotions correspondingly. Way forward is to develop more sophisticated and precise Sentiment Analysis algorithms that may capture more subjectivity, context, social and cultural variables to improve the accuracy and reliability of Sentiment Analysis in Educational Data Mining.

5.5.3 Addressing Data Imbalance and Bias

Sentiments may be expressed to add the value statements or may dominate over other less intensive emotions in the given situation. Some common imbalance issues like oversampling, undersampling, or utilizing ensemble learning approaches hinder future growth. Primary or secondary data collection processes or sentiment lexicons are required to keep as free, fair, and unbiased to address these issues and to mitigate the risk of biasness and other inherent biases.

5.5.4 Interpretable and Explainable Models [19]

The interpretability and explainability of Sentiment Analysis are required to satisfy authenticity and logical aspects. The models are to be validated before its acceptance among learners and other stakeholders. The reasons behind the sentiment prediction and related transparency are crucial to satisfying contemporary issues. This will help in adoption, adaptation and effective utilization of outputs of the Sentiment Analysis in educational decision-making processes.

5.5.5 Challenges in Scaling Sentiment Analysis to Large Educational Datasets

Sentiment Analysis has been used as a valuable tool in Educational Data Mining for understanding the emotions and opinions of learners and other stakeholders. Having with large-scale educational databases, there is a growing demand to scale up Sentiment Analysis techniques to handle the large volume and complexity data structure. However, scaling Sentiment Analysis to large educational databases poses several challenges. In this paper, we explore the key challenges on scaling up Sentiment Analysis to large educational databases and discuss potential solutions to overcome potential issues.

We can achieve optimal level moving plan by addressing the challenges related to data volume, variability, domain specificity, annotation, document length, and model interpretability. It is required for harnessing the potential of Sentiment Analysis and extracting valuable insights from educational data. By leveraging efficient algorithms, domain-specific lexicons, automated annotation techniques, and interpretability methods, Sentiment Analysis can contribute to evidence-based decision-making, personalized/recommender education, and improved learning experiences out of enormous databases.

5.5.5.1 Data Volume and Processing Efficiency

Large educational databases contain thousands or even millions of documents, e.g. learner's essays, discussion/chat forum posts, social media chats or feedback surveys. Processing such enormous text data for Sentiment Analysis becomes computationally challenge and time-consuming. Efficient algorithms and scalable techniques are needed to process the volume of data while maintaining minimum processing times. Distributed computing frameworks and parallel processing techniques may be used to accelerate processes of Sentiment Analysis on large educational databases.

5.5.5.2   Data Variability and Heterogeneity

Large educational databases are known for their variability and heterogeneity. Textual/audio/video data from different sources like courses, subjects, disciplines, applications, and educational platforms. They may exhibit variations in writing styles, language usage, activity on social media and sentiment expressions through different available platforms. This variability poses a challenge in developing robust Sentiment Analysis models that can be used for the different data sources. Building robust models for data variability and developing domain-adaptive Sentiment Analysis techniques are essential to ensure accurate Sentiment Analysis results on large educational databases to develop marking tools.

5.5.5.3   Domain-Specific Lexicon Development

Commonly used sentiment lexicons may not be useful for the domain-specific sentiments and nuances present in educational texts may not be correctly linked to related sentiments. Developing domain-specific sentiment lexicons tailor made to educational contexts is required for precisely analysis of sentiments. However, manually mapping and curating domain-specific sentiment lexicons for large educational databases are time-consuming and labour-intensive. Therefore, leveraging automated methods like distant supervision, transfer learning, or crowdsourcing can be applied in the development of holistic and contextually relevant sentiment lexicons to cater scale up demands of Sentiment Analysis in EDM.

5.5.5.4   Annotation and Labelling of Large Databases

Each document is annotated with sentiment labels as annotation and labelling of large databases and their linking with the specific sentiments are important requirement. Annotating manually may be a daunting task, requiring significant human effort and expertise. Semi-supervised or active/automated learning approaches can help optimize the annotation process by intelligently selecting representative samples like manual annotation and leveraging existing labeled data. Pre-trained models may reduce the labelling burden for scaling Sentiment Analysis to large educational databases.

5.5.5.5   Handling Long Documents and Contextual Understanding

Educational data comprise research papers or essay responses. They pose challenges in detection of sentiments. Traditional approach focussing on sentence-level sentiment classification may not be useful to find out expressions from the long documents. Developing techniques that capture the sentiment across the entire document while considering the contextual dependencies. Discourse structure is essential for calculating Sentiment Analysis accurately in educational texts. Natural language understanding techniques like document-level sentiment modelling or hierarchical Sentiment Analysis may address the challenges of handling long documents in Sentiment Analysis in case of large educational databases.

5.5.5.6 Model Interpretability and Trustworthiness

Sentiment Analysis plays a significant role in educational decision-making. It ensures the interpretability and trustworthiness of Sentiment Analysis models. Scaling Sentiment Analysis to large educational databases/datasets should not compromise the transparency and explainability of the models. Employing techniques such as model interpretability methods, Sentiment Analysis model explanations or post-hoc interpretability approaches can help generate insights. The Sentiment Analysis models do predictions, foster trust, and repose confidence in the analysis results.

5.6 Challenges in Integration of Sentiment Analysis with Other Educational Analytics [16]

The integration of Sentiment Analysis with other educational analytics holds great promise for extracting deeper insights from educational data and its mining for defined purposes. By combining Sentiment Analysis with techniques of learning analytics, social network analysis, and predictive modelling, all stakeholders may gain a comprehensive understanding of learner's sentiments, their engagement and matching prediction with the arrived learning outcomes. However, the integration of Sentiment Analysis with other educational analytics presents several challenges that can be minimized with help of finding more relevant results from the new variables and their measurements.

Integrating Sentiment Analysis with other educational analytics techniques is in fact a complex endeavour, but it offers tremendous potential for extracting meaningful insights from educational data and its mining processes. Overcoming the challenges of data integration, feature engineering, scalability, interpretability, ethical considerations, and decision support will pave the way for effective integration and utilization of Sentiment Analysis in the broader context of educational analytics. By addressing these challenges, all stakeholders can gain a comprehensive understanding of learner's sentiments, engagement, and learning outcomes, leading to data-informed decision-making, adaptive course corrections and improved educational experience and expertise.

Here, we have explained the key challenges and discussed potential strategies to overcome them for effective integration of Sentiment Analysis with other educational analytics techniques.

5.6.1   Data Integration and Compatibility

Integrating Sentiment Analysis with other educational analytics requires the integration of different data sources, subsources, and formats. Educational data often come from diverse platforms like learning management systems, social media forums, or online discussion chat history. The said data sources may have variations on account of different data formats, metadata, or sentiment labelling schemes or making data integration complex, etc. Developing data integration pipelines, standardized data formats, and interoperability frameworks may address the challenge and enabling seamless integration of Sentiment Analysis with other educational analytics.

5.6.2   Feature Engineering and Selection

Integrating Sentiment Analysis with other educational analytics necessitates the selection and combination of relevant features from different data sources to get desired output. These aspects are basic tools to have compatible combinations to give fruitful results. Each technique is based on different sentiment scores, social network metrics, or learner activity patterns. Feature engineering and proper selection are challenging aspects while dealing with multiple data sources, each with its own unique characteristics and having specific requirements. Developing feature engineering frameworks by considering the specific needs of each analytics technique and enabling feature selection processes based on their relevance and impact can be facilitated to get effective integration and further references.

5.6.3   Scalability and Performance

As educational datasets grow in volume, complexity, scalability and performance; integrating Sentiment Analysis with other analytics techniques becomes critical. Processing of large-scale datasets for Sentiment Analysis integrating with other analytical tasks can be computationally demanding and time-consuming. Employing distributed computing frameworks, parallel processing techniques and optimized algorithms can enhance the scalability and performance of Sentiment Analysis integration which ensure timely findings of insights and analysis outcomes.

5.6.4   Interpretability and Consistency

Integrating Sentiment Analysis with other educational analytics techniques requires ensuring interpretability and consistency across different analytical results. Each technique may generate its own set of metrics, models, or visualizations. It makes challenging to interpret and compare all results and its interconnections holistically. Developing visualization techniques, dashboards, or summary reports that provide a unified view of Sentiment Analysis and other educational analytics outputs can enhance interpretability and promote consistent interpretation across the different stakeholders.

5.6.5 Ethical and Privacy Considerations

Integrating Sentiment Analysis with other educational analytics techniques necessitates careful consideration of prevalent ethical and privacy concerns. Educational data often contain very sensitive information, and Sentiment Analysis may uncover personal sentiments and emotions expressed by learner or other stakeholders. It may be misused later against them. Ensuring data privacy, obtaining informed consent, and adhering to privacy and ethical guidelines are paramount when integrating Sentiment Analysis with other educational analytics techniques. Establishing robust data governance frameworks, privacy-preserving techniques, and anonymization methods can only help to address these concerns. Safeguarding the privacy and ethical use of educational data are most critical factors to be taken care of.

5.6.6 Actionable Insights and Decision Support

The goal of integrating Sentiment Analysis with other analytics techniques is to derive actionable insights and provide recommendations about decision support system to educational stakeholders. However, presenting actionable insights in a user-friendly manner and facilitating their incorporation into decision-making processes pose challenges. Developing interactive visualizations, decision support systems, or recommendation engines that leverage Sentiment Analysis outputs and combine them with other educational analytics results will enhance the usability and impact of integrated analytics solutions.

5.6.7 Privacy and Ethical Concerns in Sentiment Analysis in Educational Setting [20–22]

Sentiment Analysis in educational settings has gained momentum in recent past and drawn significant attention due to its potential to understand stakeholder's emotions, engagement, and finally their well-being. However, as Sentiment Analysis involves analysing personal textual data, privacy and ethical considerations, it becomes paramount to protect its integrity. It is crucial to address the privacy and ethical concerns associated with Sentiment Analysis in educational settings to ensure the responsible and ethical use of stakeholder's data.

Privacy and ethical concerns are very critical aspects in Sentiment Analysis within educational settings. Striking a balance between leveraging Sentiment Analysis to gain valuable insights and protecting stakeholder's privacy is utmost necessity. By implementing robust privacy protocols, addressing biases, considering real-life contexts, and promoting transparency and ethical use, Sentiment Analysis becomes a valuable tool for improving educational experiences while respecting the rights and well-being of all individuals. It is essential for educational institutions and all other stakeholders to proactively address these concerns to foster a trustworthy and responsible Sentiment Analysis framework in the educational setting to improvise educational system.

The key privacy and ethical concerns that arise after integration of other educational analytics with Sentiment Analysis are discussed as undernoted along with the mitigation strategies.

5.6.7.1 Data Privacy and Consent

One of the primary concerns in Sentiment Analysis is safeguarding the privacy of learners' data. Educational data like essays, discussion forum chats or posts, or social media interactions, may contain sensitive and personal information. That may be misused later. It is essential to establish robust data privacy protocols and frameworks that guide and govern the collection, storage, and usage of learners' data. Obtaining informed consent from them and ensuring transparency about data collection and analysis processes are vital steps to protect their confidentiality and privacy rights.

5.6.7.2 Anonymization and De-identification

To protect learner's privacy, Sentiment Analysis should incorporate anonymization and de-identification techniques to use the analysis without getting individual information. This involves removing or obfuscating personally identifiable information from the data, e.g. names, addresses, or IDs, etc. Applying techniques like data masking, tokenization, or differential privacy which can help to preserve the anonymity of learners while still deciphering meaningful outcomes of Sentiment Analysis. It is crucial to implement robust anonymization protocols to prevent re-identification of individuals whose data have been used to arrive Sentiment Analysis results.

5.6.7.3   Fairness and Bias

Sentiment Analysis models may exhibit biases which lead to unfair treatment or discrimination against certain individuals or groups. Biases can arise from biased training data, preconceived notions, or the subjective nature of Sentiment Analysis. It is essential to address biases and ensure fairness in Sentiment Analysis by carefully crafting guidelines and curating training datasets, employing unbiased sentiment lexicons and regularly evaluating and auditing the performance of Sentiment Analysis models. Striving for fairness and transparency should be an integral part of the Sentiment Analysis process in the educational settings.

5.6.7.4   Contextual Understanding and Misinterpretation

Sentiment Analysis in Educational Data Mining and their linked settings needs to consider the nuances or critical factors of the contexts and avoid misinterpretation of sentiments. Educational texts often contain complex language, sarcasm, irony, or cultural references that can lead to incorrect Sentiment Analysis results. Failing to understand the contexts correctly may result in misrepresentation or misjudgement of learner's sentiments. Employing sophisticated Sentiment Analysis algorithms that account for context, domain knowledge, and cultural variations can help to mitigate the risk of misinterpretation and improve the accuracy of the Sentiment Analysis.

5.6.7.5   Ethical Use of Sentiment Analysis Results

Ethical considerations extend beyond the data privacy and fairness to the responsible use of Sentiment Analysis results to protect individual's identity and privacy. Educational organization and all other stakeholders must ensure that Sentiment Analysis outputs are used ethically and responsibly for the benefits of the individuals or references may be used for enhancing efficiency and effectiveness of the education system. The insights derived from Sentiment Analysis should be solely used to improvise educational experiences, personalize interventions and recommendations, and overall enhance learner's well-being. It is crucial to establish clear guidelines and establish ethical frameworks that govern the utilization of Sentiment Analysis results, preventing its misuse or harmful practices that may impact the concerned stakeholder negatively.

5.6.7.6   Transparency and Stakeholder Awareness

Maintaining transparency and promoting stakeholder awareness are crucial factors in addressing privacy and ethical concerns in Sentiment Analysis. Educational institutions should communicate openly with parents, learners and educators about the purpose, methods, and significant impact of Sentiment Analysis. Clear policies, consent mechanisms, and guidelines should be incorporate and established to ensure that stakeholders have a comprehensive understanding of the Sentiment Analysis process, their rights, and the safeguards in place to protect their privacy in the domain of Educational Data Mining.

5.7 Conclusion

The emerging trends and innovations in educational Sentiment Analysis present significant potential for improving experience and educational practices. Sentiment Analysis, Contextual Analysis, Real-Time Analysis, Multimodal Explainable Sentiment Analysis, Social Network Analysis, and Domain-Specific Sentiment Lexicons are reshaping the landscape of Sentiment Analysis in education. By adopting these advancements, administrator, educators, other stakeholder and policymakers can gain valuable insights into learner sentiments, personalize interventions, foster engagement, and ultimately enhance the overall educational path or journey for learners. The future of Sentiment Analysis in Educational Data Mining holds immense potential for transforming education by providing valuable insights into learner sentiments, emotions, and attitudes. Fine-grained Sentiment Analysis, cross-domain analysis, emotion recognition, and multilingual analysis will enable a deeper understanding of sentiments expressed in diverse educational contexts. However, challenges related to data privacy, data imbalance, subjectivity and interpretability must be addressed for Sentiment Analysis to reach its full potential in education. By actively addressing these challenges and embracing future directions, Sentiment Analysis in Educational Data Mining will contribute to personalized learning, enhanced learner engagement, and improved educational experiences. With continued research, innovation, and collaboration, Sentiment Analysis and Educational Data Mining can shape the future of education and pave the way for a more effective and engaging learning experience.

References

1. Zhou J, Ye J-M (2020) Sentiment analysis in education research: a review of journal publications. Interact Learn Environ. https://doi.org/10.1080/10494820.2020.1826985
2. Shaik T, Tao X, Dann C, Xie H, Li Y, Galligan L (2023) Sentiment analysis and opinion mining on educational data: a survey. Nat Lang Process J 2:100003. https://doi.org/10.1016/j.nlp.2022.100003
3. Wankhade M, Rao ACS, Kulkarni C (2022) A survey on sentiment analysis methods, applications, and challenges. Artif Intell Rev 55(7):5731–5780. https://doi.org/10.1007/S10462-022-10144-1
4. Liang W et al (2022) Advances, challenges and opportunities in creating data for trustworthy AI. Nat Mach Intell 4(8):669–677. https://doi.org/10.1038/s42256-022-00516-1
5. Zhou J, Ye J-M (2023) Sentiment analysis in education research: a review of journal publications. Interact Learn Environ. https://doi.org/10.1080/10494820.2020.1826985

6. Evans D (2002) Emotion: the science of sentiment. Am J Orthopsychiatry 72(4). https://doi.org/10.1037//0002-9432.72.4.601
7. Hovy EH (2015) What are sentiment, affect, and emotion? Applying the methodology of Michael Zock to sentiment analysis, pp 13–24. https://doi.org/10.1007/978-3-319-08043-7_2
8. Hajrizi R, Nuçi KP (2020) Aspect-based sentiment analysis in education domain, Oct 2020. [Online]. Available: http://arxiv.org/abs/2010.01429
9. Dashtipour K et al (2016) Multilingual sentiment analysis: state of the art and independent comparison of techniques. Cogn Comput 8(4):757–771. https://doi.org/10.1007/S12559-016-9415-7/TABLES/2
10. Breitung C, Kruthof G, Müller S (2023) Contextualized sentiment analysis using large language models. SSRN Electron J. https://doi.org/10.2139/SSRN.4615038
11. Altrabsheh N, Gaber MM, Cocea M (2013) SA-E: sentiment analysis for education. Front Artif Intell Appl 255:353–362. https://doi.org/10.3233/978-1-61499-264-6-353
12. Fiok K, Farahani FV, Karwowski W, Ahram T (2022) Explainable artificial intelligence for education and training. J Def Model Simul 19(2):133–144. https://doi.org/10.1177/154851292 11028651
13. Cavalcanti AP, Mello RF, Gašević D, Freitas F (2023) Towards explainable prediction feedback messages using BERT. Int J Artif Intell Educ. https://doi.org/10.1007/S40593-023-00375-W
14. Hasib K, Rahman F, Hasnat R (2022) A machine learning and explainable AI approach for predicting secondary school student performance. In: 2022 IEEE 12th annual computing and communication workshop and conference. Accessed 27 Jan 2024. [Online]. Available: https://ieeexplore.ieee.org/abstract/document/9720806/?casa_token=bqv-ol8EoecAAAAA:o78Ad7 z2RUhs4SaNmkUuQ9RCebA7LpGRawiSjR_zcHdFXzdLf35SXTBF3DU0aq4tYGLTxm-1gvWm3REv
15. Prasad MSVCR, Mohammed M, Dhanush G, Vamsi A, Lakshmi Prasanna T (2020) Lexicon based sentiment analysis of tweets. Int J Adv Sci Technol 29(3)
16. Romero C, Ventura S (2020) Educational data mining and learning analytics: an updated survey. Wiley Interdiscip Rev Data Min Knowl Discov 10(3). https://doi.org/10.1002/widm.1355
17. Shaik T, Tao X, Li Y, Dann C, McDonald J (2022) A review of the trends and challenges in adopting natural language processing methods for education feedback analysis. IEEE Access. Accessed 27 Jan 2024. [Online]. Available: https://ieeexplore.ieee.org/abstract/document/978 1308/
18. Zhang W, Li X, Deng Y, Bing L, Lam W (2023) A survey on aspect-based sentiment analysis: tasks, methods, and challenges. IEEE Trans Knowl Data Eng 35(11):11019–11038. https://doi.org/10.1109/TKDE.2022.3230975
19. Fiok K, Farahani F, Karwowski W, Ahram T (2022) Explainable artificial intelligence for education and training. J Def Model Simul. Accessed 27 Jan 2024. [Online]. Available: https://doi.org/10.1177/15485129211028651
20. Šlibar B, Zlatić L, Ređep N (2021) Ethical and privacy issues of learning analytics in higher education. In: ICERI2021 proceedings. Accessed 27 Jan 2024. [Online]. Available: https://library.iated.org/view/SLIBAR2021ETH
21. Sabourin J, Kosturko L, Fitzgerald C, Mcquiggan S (2015) Student privacy and educational data mining: perspectives from industry. Accessed 27 Jan 2024. [Online]. Available: https://ora.ox.ac.uk/objects/uuid:1d91daf7-7dfa-41ac-bb84-1bcf87cb6ea8
22. Prinsloo P, Slade S, Khalil M (2019) Student data privacy in MOOCs: a sentiment analysis. Dist Educ. Accessed 27 Jan 2024. [Online]. Available: https://doi.org/10.1080/01587919.2019.163 2171